ACIDS AND BASES

Solvent Effects on Acid–Base Strength

Acids and Bases

Solvent Effects on Acid–Base Strength

BRIAN G. COX
AstraZeneca, Macclesfield, UK

UNIVERSITY PRESS

Great Clarendon Street, Oxford, OX2 6DP,
United Kingdom

Oxford University Press is a department of the University of Oxford.
It furthers the University's objective of excellence in research, scholarship,
and education by publishing worldwide. Oxford is a registered trade mark of
Oxford University Press in the UK and in certain other countries

© Brian G. Cox 2013

The moral rights of the author have been asserted

First Edition published in 2013

Impression: 1

All rights reserved. No part of this publication may be reproduced, stored in
a retrieval system, or transmitted, in any form or by any means, without the
prior permission in writing of Oxford University Press, or as expressly permitted
by law, by licence or under terms agreed with the appropriate reprographics
rights organization. Enquiries concerning reproduction outside the scope of the
above should be sent to the Rights Department, Oxford University Press, at the
address above

You must not circulate this work in any other form
and you must impose this same condition on any acquirer

British Library Cataloguing in Publication Data

Data available

ISBN 978–0–19–967051–2 (hbk.)
      978–0–19–967052–9 (pbk.)

Printed and bound by
CPI Group (UK) Ltd, Croydon, CR0 4YY

Links to third party websites are provided by Oxford in good faith and
for information only. Oxford disclaims any responsibility for the materials
contained in any third party website referenced in this work.

Preface

The activation of organic molecules by acids and bases, through addition or removal of a proton, and the catalytic reactions that ensue are of fundamental importance in organic chemistry. Many other processes, including the selective isolation of products by extraction or crystallization, chromatographic separation of ionizable compounds, and the formulation of pharmaceutical and agrochemical actives through salt formation, are also intimately linked to acid–base (protolytic) equilibria.

Current understanding of protolytic equilibria derives largely from, and relates primarily to, studies in aqueous solution, and indeed there is no other class of reaction for which such accurate data are available. In practice, however, perhaps the majority of chemical reactions, especially in synthetic chemistry, but also in many other technologies, such as lithium batteries, involve the use of non-aqueous or mixed-aqueous media, and there is an increasing focus in both academic and industrial laboratories on quantitative understanding and modelling of such processes. The solvent commonly has a profound effect on solution equilibria and the often dramatic changes to acid–base behaviour in non-aqueous solvents compared with water are in the main poorly documented and not widely understood nor appreciated.

This book seeks to remedy this deficiency by reviewing acid–base behaviour in non-aqueous solvents and providing an understanding of the results in terms of the interactions between the species involved and the solvent. The approach is primarily to relate the observed behaviour to that in aqueous solution, but importantly also to consider correlations and contrasts directly between different classes of non-aqueous solvents.

Chapters 1–3 provide the necessary background material: the definition of acids and bases and the quantitative relationship between solution pH, acid strength and species distribution; the influence of molecular structure on acid–base properties; the range of acidity in aqueous solution for different classes of acids; the properties of solvents, especially with respect to their ability to participate in hydrogen-bond formation and to donate or accept electrons; and the magnitude of the interaction energies between solvent molecules and (especially) ions, which provide the basis for an understanding of the solvent effects on the equilibria.

Chapter 4 discusses pH-scales in aqueous and non-aqueous media and describes experimental methods for determining dissociation constants (pK_a-values) in non-aqueous media and the autoionization constants of the solvents. A distinctive feature of acid–base chemistry in non-aqueous solvents is the accompanying presence of ion-pair and hydrogen-bond association equilibria, such as that between carboxylic acids and carboxylate ions, which can strongly influence dissociation of acids and bases.

Chapters 5–7 present and analyse dissociation constants in three distinct classes of solvent: *protic* solvents, such as the alcohols, which are strong hydrogen-bond donors; *basic, polar aprotic* solvents, such as dimethylformamide and dimethylsulphoxide; and *low-basicity* and *low-polarity aprotic* solvents, such as acetonitrile and tetrahydrofuran. The distinction between high-basicity and low-basicity aprotic solvents is necessary because of the dominant influence of the solvation of the proton in determining the response of dissociation constants to solvent change. Aprotic solvents are at best very weak hydrogen-bond donors, but protic solvents, by contrast, are characterized by their ability to stabilize anions by hydrogen-bond donation.

Chapter 8 describes the effect of solvent on proton-transfer equilibria, such as those involved in salt formation and in the activation of organic molecules by deprotonation. The distinction between high- and low- basicity aprotic solvents is no longer relevant because the solvated proton is not involved, but at solution concentrations typical of many practical applications, the systems often become dominated by accompanying equilibria, such as ion-pair formation. In particular, in solvents of very low polarity, whilst there is no measurable tendency to form free ions, proton-transfer to form ion-paired species is often extensive.

The book can be read at several levels. The background required for a full appreciation of the material presented is primarily a familiarity with basic-solution physical chemistry and in particular the relationship between equilibrium constants and free energy change. The important messages arising from the data and correlations presented in Chapters 5–7 can, however, essentially be derived independently of the remaining material; indeed undergraduate chemistry courses would benefit strongly from their inclusion, along with much of the material in Chapter 2. Quantitative modelling of reaction and isolation processes coupled to acid–base equilibria is perhaps best left to more specialist groups, such as process engineers or those involved in the study of reaction mechanisms.

Acknowledgements

My ongoing interest in, and enthusiasm for, solution kinetics and thermodynamics was sparked by three people: P. T. McTigue (University of Melbourne); R. P. Bell (Universities of Oxford and Stirling); and A. J. Parker (Australian National University). More recently, my approach to chemistry has been strongly influenced by numerous colleagues in ICI and its daughter companies, Syngenta and AstraZeneca, but especially by John H. Atherton, Lai C. Chan, and Euan A. Arnott, to whom special thanks are due.

I would also like to thank AstraZeneca for use of the facilities that made writing of this book possible, and for the provision of a stimulating working atmosphere; in the latter context, I mention in particular Andy Jones.

Contents

1 Introduction — 1
   1.1 Why are non-aqueous solvents important? — 3
      1.1.1 Range of accessible pH-values — 3
      1.1.2 Reactivity of acids and bases — 5
      1.1.3 Ionizable compounds and high-performance liquid chromatography (HPLC) — 5
   1.2 Classification of solvents and their properties — 5
      1.2.1 Electron donor–acceptor properties of solvents — 6
   References — 9

2 Acid–Base Equilibria: Quantitative Treatment — 10
   2.1 Definitions — 10
   2.2 pH and acid–base ratios — 11
   2.3 pH-dependence of species distribution — 12
   2.4 Acid strengths and molecular structure: dissociation constants in water — 15
      2.4.1 The strength of the X–H bond — 15
      2.4.2 Charge dispersion on anions, A^-, or cations, BH^+ — 16
      2.4.3 The nature of X in R–X–H — 17
   2.5 Carbon acids — 17
   2.6 Acid–base equilibria — 19
   Appendix 2.1 Species distribution in acid-base systems — 19
   References — 20

3 Solvation and Acid–Base Strength — 21
   3.1 Solvation and acid dissociation constants: free energies of transfer — 21
      3.1.1 Solvent-transfer activity coefficients — 23
   3.2 Determination of free energies of transfer — 23
      3.2.1 Non-electrolytes — 23
      3.2.2 Electrolytes — 25
   3.3 Free energies of ion solvation — 27
      3.3.1 Hydration of ions — 27
      3.3.2 Solvation in pure solvents — 28
      3.3.3 Solvation in mixed solvents — 31
   3.4 Solvation of non-electrolytes — 32

3.5	Solvation energies and solvent properties	33
3.6	Solvation and acid strength	34
3.7	Summary	35
Appendix 3.1 Composition of mixed solvents		36
References		38

4 Determination of Dissociation Constants — 39

4.1 pH-scales — 39
    4.1.1 pH in aqueous media — 39
    4.1.2 pH in non-aqueous media — 41
4.2 Influence of solution concentration: activity coefficients — 41
4.3 Ion association — 43
4.4 Homohydrogen-bond formation — 44
4.5 Experimental methods for the determination of dissociation constants — 46
    4.5.1 Potentiometric titration using a glass electrode — 47
    4.5.2 Acid–base indicators — 50
4.6 Autoionization constants of solvents — 53
Appendix 4.1 Dissociation of acetic acid in the presence of sodium chloride — 54
Appendix 4.2 Ion-pair formation and pK_a-determination — 54
Appendix 4.3 Determination of homohydrogen-bond association constants, K_{AHA} — 55
References — 57

5 Protic Solvents — 59

5.1 Autoionization constants — 59
5.2 Methanol — 60
    5.2.1 Neutral acids: carboxylic acids, phenols — 61
    5.2.2 Cationic acids: protonated anilines, amines, N-heterocycles — 64
    5.2.3 Summary — 66
5.3 Higher alcohols — 66
5.4 Alcohol–water mixtures — 68
5.5 Salt formation in alcohols and aqueous–alcohol mixtures — 70
5.6 Formamide, acetamide, N-methylpropionamide — 72
5.7 Formic acid — 74
References — 75

6 High-Basicity Polar Aprotic Solvents — 76

6.1 Dimethylsulphoxide — 77
    6.1.1 Neutral acids: carboxylic acids, phenols and thiophenols, water and methanol, anilines and amides, carbon acids — 78
    6.1.2 Cationic acids (neutral bases) — 86
    6.1.3 Amino acids — 87

6.2	*N*-methylpyrrolidin-2-one, *N*, *N*-dimethylformamide, *N*, *N*-dimethylacetamide	89
	6.2.1 Neutral acids	89
	6.2.2 Cationic acids (neutral bases)	92
6.3	Liquid ammonia	93
6.4	Summary	95
6.5	Estimation of dissociation constants in basic aprotic solvents	96
References		97

7 Low-Basicity and Low-Polarity Aprotic Solvents — 99

7.1	Acetonitrile	100
	7.1.1 Neutral acids: carboxylic acids and phenols, carbon acids	100
	7.1.2 Cationic acids (neutral bases)	105
7.2	Propylene carbonate, sulpholane, acetone, methyl *iso*-butyl ketone, nitrobenzene	108
7.3	Tetrahydrofuran	111
References		115

8 Acid–Base Equilibria and Salt Formation — 117

8.1	Charge-neutral equilibria	118
8.2	Charge-forming equilibria	120
	8.2.1 Alcohols and mixed-aqueous solvents	120
	8.2.2 Polar aprotic solvents	124
	8.2.3 Non-polar aprotic solvents	127
References		128

9 Appendices: Dissociation Constants in Methanol and Aprotic Solvents — 130

9.1	Methanol	130
	9.1.1 Carboxylic acids and phenols	130
	9.1.2 Protonated nitrogen bases	131
9.2	Dimethylsulphoxide	132
	9.2.1 Carboxylic acids, alcohols, phenols	132
	9.2.2 Inorganic acids and miscellaneous	133
	9.2.3 Anilines, anilides, amides (N–H-ionization)	133
	9.2.4 Carbon acids: ketones, esters, nitroalkanes	133
	9.2.5 Carbon acids: nitriles, sulphones	134
	9.2.6 Carbon acids: fluorenes	134
	9.2.7 Cationic acids: anilinium, ammonium, pyridinium ions	135
9.3	*N*, *N*-Dimethylformamide	135
	9.3.1 Neutral acids	135
	9.3.2 Cationic acids	136
9.4	Acetonitrile	136
	9.4.1 Neutral acids	136

	9.4.2	Inorganic acids and miscellaneous	137
	9.4.3	Cationic acids: ammonium, anilinium, pyridinium ions	138
	9.4.4	Phosphazene bases	138
9.5	Tetrahydrofuran	140
	9.5.1	Neutral acids	140
	9.5.2	Cationic acids	140

Index   141

Introduction

1

When we raise the topic of solvent effects on acid–base equilibria, the underlying question is often: How does the dissociation of acids in non-aqueous media compare with that in water (with which we are broadly familiar)? The answer is that it frequently differs profoundly, but fortunately in a manner that is both rational and largely predictable.

Acid–base equilibria are intimately coupled to many of the processes involved in synthetic and analytical chemistry, including reaction rates and selectivity, solubility equilibria, partition equilibria, catalytic cycles and chromatographic retention times. Furthermore, mechanistic investigations and the development of manufacturing processes rely increasingly on quantitative modelling of the reactions and equilibria involved, which are often highly sensitive to solvent.

Some examples serve to illustrate the changes which can occur on solvent transfer. Thus, the equilibrium constant, K_e, for the reaction between acetic acid and triethylamine, eq. (1.1), in the commonly used polar aprotic solvent, N-methylpyrolidin-2-one (NMP), is smaller by *10 orders of magnitude* than that in water; proton transfer from acetic to triethylamine in aqueous solution is almost quantitative ($K_e = 10^6$), whereas in NMP the neutral species are overwhelmingly favoured ($K_e < 10^{-4}$).*

*It will be seen also that hydrogen-bond association, such as between CH_3CO_2H and $CH_3CO_2^-$, and ion-pair formation also play an important role at higher concentrations

$$CH_3CO_2H + Et_3N \underset{NMP}{\overset{H_2O}{\rightleftarrows}} CH_3CO_2^- + Et_3NH^+ \quad (1.1)$$

A related effect may be seen in the crystallization of anthranilic acid which produces different polymorphs on isolation from ethanol and ethanol–water mixtures. In solution the acid exists as a solvent-dependent, equilibrium mixture of neutral and zwitterionic forms, eq. (1.2); the stable polymorph in the solid state comprises a 1:1 mixture of these two forms.

$$\text{(1.2)}$$

Crystallization from ethanol gives an unstable polymorph containing only the neutral species, which slowly reverts to the stable polymorph on prolonged contact with the solution. Direct crystallization of the acid as the stable polymorph, however, is achieved by simple change of solvent to a mixture of ethanol and water. This occurs because of a change in the equilibrium ratio of the two forms. In ethanol the neutral species is the dominant form of anthranilic acid, and its crystallization is kinetically favoured. Increasing the water content of the solvent causes a shift in the equilibrium towards a higher proportion of the zwitterionic form, which then allows direct crystallization of the thermodynamically more stable 1:1 mixture [1].

In other cases, the stoichiometry of salts may depend upon the solvent from which they are crystallized. This is illustrated for the anilinoquinazoline kinase inhibitor, (I) [2], which may be isolated as a fumarate salt, Scheme 1.1.

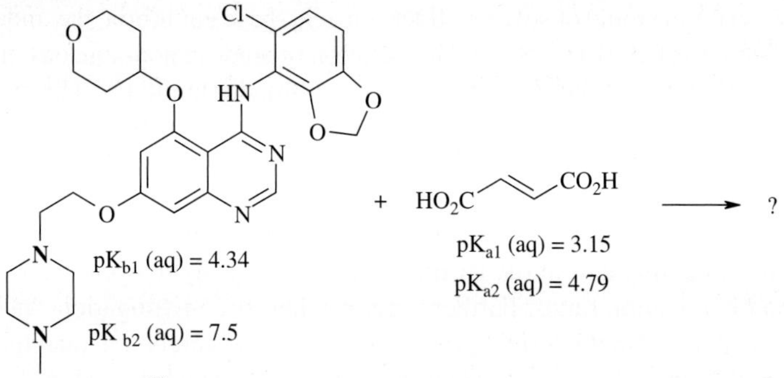

Scheme 1.1.
Isolation of the anilinoquinazoline kinase inhibitor, (I), as its fumarate salt

Crystallization from alcohol–water mixtures gives the (2:1) salt: $[(IH_2^{2+})(HA^-)_2]$ in which HA^- represents the monoanion of fumaric acid and IH_2^{2+} the diprotonated quinazoline. In water, however, the recovered product is a 1.5:1 salt, $[(HA^-)(IH_2^{2+})(A^{2-})(IH_2^{2+})(HA^-)]$, isolated as the hydrate. The most notable difference between the two salts is the presence of the difumarate anion in the latter but not the former, the equilibrium level of which is severely reduced as the alcohol content of the solvent increases due to the decreased acidity of both fumaric acid and its mono-anion.

The dominant use of, and familiarity with, water as solvent in quantitative studies of acid–base properties, has tended to obscure the importance of the influence of solvation, i.e., the extent to which the relative values of acid/base strengths in water depend upon its peculiarly strong ability to solvate both cations and anions. It is in many ways unique: among other features, it has a high dielectric constant ($\varepsilon_r = 78.5$ at 25°C), it is a strong hydrogen-bond donor and acceptor, and it is amphoteric, possessing both acidic and basic properties.

A consequence of this ability to solvate effectively both anions and cations is that pK_a-scales in water are quite different from those observed in other solvents, and this has important implications in a variety of situations: for example, synthetic reactions involving acids, bases and nucleophiles; salt formation between carboxylic acids and many pharmaceutical actives; and zwitter-ion formation in amino acids.

1.1 Why are non-aqueous solvents important?

The majority of practical syntheses are carried out in non-aqueous media. This can be for a variety of reasons: increased solubility of organic reagents and hence increased productivity; the ability to tolerate water-sensitive reagents; increased reactivity and, in some cases, more favourable selectivity. Very important also is the low tendency of most non-aqueous solvents to self-ionize, which enables them to tolerate very strong bases. Other areas, such as non-aqueous battery technology, are also assuming great importance.

1.1.1 Range of accessible pH-values

A significant restriction to the use of aqueous chemistry in synthetic applications is the limitation imposed by the ionization of water by strong bases; this severely limits the attainable pH-values, and hence the extent of ionization of weakly acidic substrates that may be achieved. The ionization of water is represented by the autoprotolysis constant (ionic product), K_w, defined by eq. (1.3):

$$K_w = [H^+][OH^-] \quad (1.3)$$

This product has a value of $1.0 \times 10^{-14} M^2$ at 25°C, which means that pH = 14 is the approximate upper limit to the pH that can be achieved in water at 25°C.* Stronger bases are immediately protonated by the water molecules, leaving only the relatively weakly basic hydroxide ion.

By contrast, aprotic solvents, such as dimethylsulphoxide (DMSO) and acetonitrile, have very low ionic products. The ionic product of DMSO, K_{DMSO}, eq. (1.4), for example, has a value of $10^{-35} M^2$ [3].

$$K_{DMSO} = [(CH_3)_2SOH^+][CH_3SOCH_2^-] \quad (1.4)$$

Thus, use of a suitably strong base, such as phosphazene base P_1 ($pK_a = 28.4$ in acetonitrile) [4], allows facile generation of sufficient concentrations of the anions of most of the weak carbon acids, such as ketones, for successful use in syntheses in aprotic solvents, Scheme 1.2 [5].

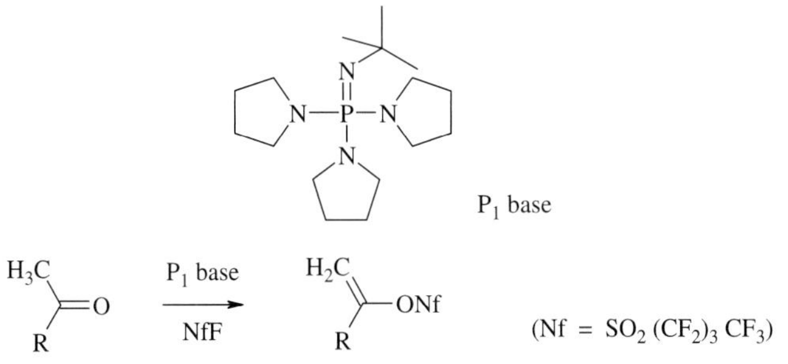

*pH 14 corresponds approximately to [NaOH] = 1M; values of around pH 15 are possible in very concentrated solutions of NaOH

Scheme 1.2.
Generation of enols from ketones in dimethylformamide

A similar limitation arises during attempts to generate strongly *acidic* solutions in water, in order to promote, for example, the nitration or sulphonation of unreactive aromatic substrates. The acidity of the solutions is limited by the

*Quantitative aspects of acids and bases are detailed in Chapter 2

basicity of water, with the formal pK_a of the hydronium ion (H_3O^+) being given by eq. (1.5).*

$$K_a(H_3O^+) = [H_2O] = 55.5\,M; \quad pK_a(H_3O^+) = -1.7 \qquad (1.5)$$

The consequence is that any acid with a pK_a lower than that of H_3O^+ (which includes the majority of mineral acids) will simply be converted to H_3O^+ and the corresponding acid anion by protonation of water.

Very high acidities can, however, be generated in solvent mixtures of water and acids such as sulphuric acid. The properties of concentrated acid solutions differ considerably from those in dilute solutions, with the acidity (as measured by the ability to protonate indicators) increasing much more rapidly than the concentration [6, 7]. At high acid concentrations there is insufficient water available to solvate the proton effectively and hence the proton activity (acidity) increases markedly. The increases in acidity can amount to many orders of magnitude, and have been expressed quantitatively in terms of an *acidity function*, H_o, which represents the effective pH in these concentrated acid solutions. This topic will not be considered further in this book.

Similar considerations, involving either ionization or protonation of the solvent, apply to a number of other solvent systems. Acetic acid as solvent, for example, can support very strong acids, and indeed it is a common solvent for acids such as HBr, but strong bases are quantitatively protonated to give the acetate salt of the protonated base. The reverse is true for ammonia, which is protonated by even relatively weak acids, but can support very strong bases, because of its reluctance to form NH_2^-. Fig. 1.1 illustrates approximate acid/base ranges for some representative solvents.

Solvents of very low polarity and having no ionizable protons, such as cyclohexane, are in principle able to tolerate a wide range of acids and bases, but their usefulness as reaction media is of course restricted by their poor ability to dissolve ionic species.

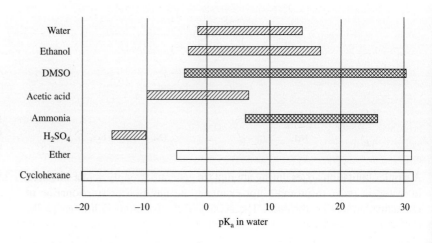

Fig. 1.1.
Range of existence of acids and bases in different solvents

1.1.2 Reactivity of acids and bases

The transfer of acids and bases to non-aqueous media not only allows a wider range of basic species to exist at higher concentrations, but, perhaps of greater practical importance, it also alters their reactivity. One consequence is that relative and absolute acid–base strengths are frequently altered drastically. This was illustrated earlier for the reaction between triethylamine and acetic acid, eq. (1.1), the equilibrium constant for which decreases by 10 orders of magnitude on transfer from N-methylpyrollidin-2-one (NMP) to water. Similar behaviour is also observed in other solvents, such as dimethylformamide and dimethylsulphoxide.

Furthermore, the reactivity of anionic bases and nucleophiles in solvents such as dimethylformamide and dimethylsulphoxide is greatly increased relative to that in water, often with great benefit to the rates and yields of synthetic reactions.

1.1.3 Ionizable compounds and high-performance liquid chromatography (HPLC)

The pH of mobile phases used in liquid chromatography, typically acetonitrile-water, tetrahydrofuran-water, or *iso*-propanol-water mixtures [8–10], is an area of importance in the separation of acid- or base-sensitive substrates. The pH affects the degree of ionization of acidic or basic substrates, which in turn affects their retention times. In particular, when the pH is close to the pK_a of the analyte, slight variations of pH may cause notable changes in the retention times. This has led to a considerable amount of work on measuring acid–base dissociation constants in mixed-aqueous solvents.

An understanding, both qualitative and quantitative, of the properties of acids and bases in different solvents is therefore of some general significance, and it is the purpose of this book to review acid–base behaviour in non-aqueous solution and to provide an understanding of the results in terms of the interactions between the species involved and the solvent.

1.2 Classification of solvents and their properties

Protic solvents, such as water, alcohols, formamide, and formic acid, are strong hydrogen-bond donors. *Polar aprotic* solvents are no more than very weak hydrogen-bond donors [11].

Behind this classification lies the simple guiding rule that solvents with hydrogen bound only to carbon are normally very weakly acidic, are at best poor hydrogen-bond donors, and exchange very slowly, if at all, with D_2O. In contrast, solvents with hydrogen bound directly to electronegative atoms, such as oxygen and nitrogen, exchange rapidly with D_2O and form strong hydrogen bonds with suitable acceptors.

Common polar aprotic solvents are dimethylformamide (DMF), N-methylpyrrolidin-2-one (NMP), dimethylsulphoxide (DMSO), dimethylacetamide (DMAC), acetonitrile, propylene carbonate (PC), sulpholane (TMS), acetone, nitromethane, and nitrobenzene. Perhaps surprisingly, liquid ammonia is also a very poor hydrogen-bond donor and acts as a typical, but strongly basic, aprotic solvent, despite having protons bound to nitrogen (Chapter 6).

Low-dielectric solvents ($\varepsilon_r < \sim 15$) are more problematic with respect to quantitative treatment of ionic equilibria, because of the difficulty in observing the properties of solvent-separated ions in such media, but are used extensively in synthetic processes.

This distinction between protic and polar aprotic solvents was first recognized in response to the observed marked acceleration in common organic reactions, such as substitution, elimination, addition and abstraction, involving anionic nucleophiles or bases, eq. (1.6), when carried out in polar aprotic solvents compared with those in protic solvents [11].

$$Nu^- + RX \rightarrow \text{products of substitution, elimination,}$$
$$\text{addition, or abstraction} \quad (1.6)$$

The resultant increased yields, increased selectivity, and decreased reaction times have led to the widespread use of polar aprotic solvents in organic synthesis at both laboratory and industrial scales.

The classification of solvents as protic or polar aprotic serves the useful purpose of highlighting the importance of hydrogen-bonding interactions in acid–base behaviour, particularly for neutral acids, such as carboxylic acids and phenols, and the reactivity of anionic nucleophiles, but there are, of course, other types of interaction, such as dispersion force, ion–dipole, and dipole–dipole interactions, which must also be considered when analysing the influence of solvent on acid–base behaviour.

Several parameters have been introduced to model the interactions between solutes and solvents, especially with respect to the ability of solvents to stabilize cations and anions. Reichardt [12] has discussed this topic extensively, and some selected properties of common protic and polar aprotic solvents are listed in Table 1.1.

1.2.1 Electron donor–acceptor properties of solvents

Thermodynamic studies of the interactions between ions and solvent molecules in the gas phase and in solution (see Chapter 3, Section 3.3) suggest strongly that the solution properties of ions are dominated by specific interactions occurring primarily *within the first solvation sphere*. It is, therefore, to be anticipated that those solvent properties or parameters that provide some measure of the solvent molecule's ability to donate or accept electron pairs will be of most use in analysing solvent effects on acid–base equilibria.

The *dielectric constant*, ε_r, is a bulk-solvent property that can, over a wide range of solvents, give a general indication of the ability to support ions. Among the relatively polar solvents in which we are primarily interested, however, it has little useful predictive power with respect to solvating ability; thus, for example, MeOH, DMF and NMP have similar dielectric constants but differ widely as reaction media. It does, however, provide a good indication

Table 1.1 Properties of solvents[a]

Solvent[b]	Bp/°C	Dielectric constant ε_r	Dipole moment μ/D[c]	Polarization/cm^{3d}	Donor Number[e] DN	H-bond basicity[e] β	Acceptor Number[e] AN
H$_2$O	100.0	78.5	1.84	17.3	18	0.47	54.8
MeOH	64.5	32.6	1.70	36.9	19	0.66	41.5
EtOH	78.3	24.6	1.69	51.8	18.5	0.75	37.1
i-PrOH	82.4	19.9	1.65	65.2		0.84	33.7
HCONH$_2$	193	109.5	3.25	38.8	24.7	0.48	41.5
DMF	152.5	36.7	3.82	71.3	26.6	0.69	16.0
DMAC	165.5	37.8	3.79	85.2	27.8	0.76	13.6
NMP	202	31.5	4.09	87.2	27.3	0.77	13.3
DMSO	189	48.9	4.3	66.7	29.8	0.76	19.3
HMPA	235	29.6	5.37	158	38.8	1.05	9.8
PC	241.7	64.9	4.95	82.0	15.1	0.40	18.3
TMS	278.3	43.3	4.80	89.0	14.8	0.39	19.2
MeCN	80.1	37.5	3.84	48.4	14.1	0.40	18.9
Acetone	56.1	20.6	2.70	66.7	17.0	0.43	12.5
PhNO$_2$	210.9	34.8	4.4	94.5	4.4	0.30	14.6
CH$_3$NO$_2$	100.8	38.6	3.44	49.2	2.7	0.06	20.5
THF	66.0	7.6	1.74	55.8	20.0	0.55	8.0

[a] Ref [12], [19]; [b] Abbreviations: DMF, N,N-dimethylformamide; DMAC, dimethylacetamide; NMP, N-methylpyrrolidin-2-one; DMSO, dimethylsulfoxide; HMPA, hexamethylphosphoramide; PC, 4-methyl-1,3-dioxolane (propylene carbonate); TMS, tetrahydrothiophene 1,1-dioxide (tetramethylene sulfone, sulfolane); MeCN, acetonitrile; PhNO$_2$, nitrobenzene; CH$_3$NO$_2$, nitromethane; THF, tetrahydrofuran; [c] $1D = 3.336 \times 10^{-33}$ Cm; [d] Molar polarization, $\{(\varepsilon_r - 1)/(\varepsilon_r + 2)\}(MW/\rho)$; [e] See definitions below

of the likely extent of association between ions in solution—an area of great importance in low dielectric media.

The *dipole moment*, μ, is a molecular property, but it does not give a direct indication of the ability to solvate ions. This is because it depends upon the separation of the (partial) charges, as well as their magnitude, and a large dipole moment may be as much a reflection of the distance between the charges as of their magnitudes. It is, therefore, also of limited use as a predictor of solvating ability.

The *Donor Number*, DN, first introduced by Gutmann [13], provides a widely used and easily understood simple electron donor (Lewis basicity) scale. The Donor Number is defined as the negative enthalpy of adduct formation between the solvent molecule, S, and SbCl$_5$ in dilute solution in the non-coordinating solvent dichloromethane, eq. (1.7).

$$S + SbCl_5 \rightleftharpoons S.SbCl_5 \quad (DN = -\Delta H/\text{kcal mol}^{-1}) \quad (1.7)$$

Values are large for solvents such as DMSO, DMF, and NMP, modest for water and alcohols, and smaller for more weakly basic polar solvents, such as MeCN.

An analogous *Lewis basicity scale* for aprotic solvents, derived from the enthalpy of 1:1 adduct formation with BF$_3$ in dichloromethane, has also been reported [14] and generally correlates well with the SbCl$_5$-derived donor numbers.

It may be noted that donor numbers and related scales refer to the properties of isolated solvent molecules in inert media, and hence their relevance to the properties of bulk solvents is somewhat uncertain for the highly associated protic solvents. An estimate of bulk solvent donicities has been obtained from the solvent dependence of ^{23}Na-NMR chemical shifts [15]. These correlate very well with donor numbers for a variety of aprotic solvents, but suggest that higher donor numbers for bulk water, methanol, and ethanol, of 33, 26, and 31, respectively, may be more appropriate than those given in Table 1.1. Marcus has reviewed the topic of solvents as electron pair donors [16].

The *Acceptor Number*, AN, introduced also by Gutmann [13], complements the donor numbers by providing a measure of the electrophilic properties of solvents. It is derived from the ^{31}P-NMR chemical shift produced when Et$_3$PO is dissolved in the solvent, eq. (1.8).

$$\text{Et}_3\text{PO} + S \rightleftharpoons \overset{\delta+}{\text{Et}_3\text{PO}} \rightarrow \overset{\delta-}{S} \qquad (1.8)$$

Electron donation to the solvent from the oxygen bonded to phosphorous produces a solvent-dependent downfield shift; the shifts are scaled between zero (hexane) and 100 (Et$_3$PO – SbCl$_5$ complex in 1,2-dichloroethane). Acceptor Numbers are, as expected, highest for the protic solvents, H$_2$O to formamide (Table 1.1), because of their ability to form hydrogen bonds with Et$_3$PO.

The *electronic transitions* of various indicator molecules are strongly solvent-dependent, and the transition energies in different solvents have also been used as an empirical measure of solvent polarity. The most widely used and extensively tabulated of these parameters is the E$_T$(30) value [12], which is the lowest energy transition of the pyridinium phenol betaine, **II**, expressed in kcal mol^{-1}.

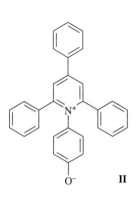

II

In addition to non-specific, dye-solvent interactions, the transition energies respond particularly to the Lewis acidity of solvents because of the relatively localised negative charge on the betaine phenolic oxygen. The probe molecule, however, does not register solvent Lewis basicity, as the positive charge of the pyridine moiety is highly delocalised. E$_T$(30) values tend, therefore, to correlate more closely with acceptor numbers.

Kamel, Taft, Abraham, and co-workers have also used spectroscopic measurements to develop a scale of *hydrogen-bond acceptor basicity*, β, and a scale of *hydrogen-bond donor acidity*, α, for use in multi-parameter correlations [17–19]. It is not possible to give a simple qualitative description of them, but they are derived from the absorption energies of selected probe solutes in the solvents in question after subtracting the effect that non-acidic and non-basic solvents would have on the same solutes. They are widely quoted in discussions of solvent basicity in particular and for this reason we also include β-values in Table 1.1. They show an almost quantitative relationship to the corresponding Donor Numbers, except for the alcohols and nitrobenzene, which exhibit relative higher basicity when measured on the β-scale.

For the remainder of this text we will use the parameters in Table 1.1 as representative and easily understood indicators of different aspects of solvent polarity.

References

[1] Black, S. N., Davey, R. J., Cox, B. G., Collier, E., Tower, C. World Congress of Chemical Engineering, 7th Glasgow, 2005, 82338/1-82338/10
[2] Hennequin, L. F., Allen, J., Breed, J., Curwen, J., Fennell, M., Green, T. P., Lambert van der Brempt, C., Morgentin, R., Norman, R. A., Olivier, A., Otterbein, L., Plé, P. A., Warin, N., Costello, G. *J. Med. Chem.*, 2006, *49*, 6465
[3] Olmstead, W. N., Margolin, Z., Bordwell, F. G. *J. Org. Chem.*, 1980, *45*, 3295
[4] Schwesinger, R., Willaredt, J., Schlemper, H., Keller, M., Schmitt, D., Fritz, H. *Chem. Ber.*, 1994, *127*, 2435
[5] Lyapkalo, I. M., Vogel, M. A. K. *Angew. Chem. Int. Ed.*, 2006, *45*, 4019
[6] Paul, M. A., Long, F. A. *Chem. Rev.*, 1957, *57*, 1
[7] Long, F. A., Paul, M. A. *Chem. Rev.*, 1957, *57*, 935
[8] Subirats, X., Bosch, E., Rosés, M. *J. Chromatogr. A*, 2009, *1216*, 2491
[9] Espinosa, S., Bosch, E., Rosés, M. *J. Chromatogr. A*, 2002, *964*, 55
[10] Bosch, E., Bou, P., Allermann, H., Rosés, M. *Anal. Chem.*, 1996, *68*, 3651
[11] Parker, A. J. *Chem. Rev.*, 1969, *69*, 1
[12] Reichardt, C. 'Solvents and Solvent Effects in Organic Chemistry', 3rd Edn. Wiley-VCH, 2004
[13] Gutmann, V. 'The Donor–Acceptor Approach to Molecular Interactions', Plenum Press, 1978
[14] Maria, P-C., Gal, J-F. *J. Phys. Chem.*, 1985, *89*, 1296
[15] Ehrlich, R. H., Roach, E., Popov, A. I. *J. Am. Chem. Soc.*, 1970, *92*, 4989
[16] Marcus, Y. *J. Soln. Chem.*, 1984, *13*, 599
[17] Abraham, M. H., Doherty, R. M., Kamlet, M. J., Taft, R. W. *Chem. Britain*, 1986, 551
[18] Abraham, M. H., Zhao, Y. H. *J. Org. Chem.*, 2004, *69*, 4677
[19] Marcus, Y. *Chem. Soc. Rev.*, 1993, *22*, 409

2 Acid–Base Equilibria: Quantitative Treatment

In this chapter we outline quantitative aspects of acid–base behaviour, which are in most cases equally applicable to aqueous and non-aqueous solutions. We also summarize the influence of molecular structure on acid strengths. This is derived principally from studies in aqueous solution, but the large body of data available serves as a useful reference point for much of the work in non-aqueous solvents discussed in the remainder of this book.

2.1 Definitions

The most useful and general definition of acids and bases comes from Brönsted and Lowry, independently in 1923, and states that: An acid is a species having a tendency to lose a proton and a base is a species having a tendency to add a proton [1]. This may be expressed by eq. (2.1), in which A and B represent a conjugate acid–base pair.

$$A^{n+} \rightleftharpoons B^{(n-1)+} + H^+ \tag{2.1}$$

For simplicity, the charges on A and B are often omitted, recognising always that A must of course be more positive than B by one unit. This simple definition contains no mention of the solvent, and the symbol H^+ in eq 2.1 represents the bare proton. In practice, however, the equilibrium as written cannot be realised in solution because of the unfeasibly high energy of the bare proton; any acid–base reaction must involve transfer of the proton between two species (one of which may be the solvent). This is the starting point for a practical definition of acid strength, applicable in all solvents.

If we combine two such equilibria for acids A_1 and A_2, e.g., CH_3CO_2H and $CH_3NH_3^+$, a general expression for all acids and bases can be written, as in eq. (2.2), the equilibrium constant for which, K_e, is given by eq. (2.3).

$$A_1 + B_2 \rightleftharpoons B_1 + A_2 \tag{2.2}$$

$$K_e = \frac{[A_2][B_1]}{[B_2][A_1]} \tag{2.3}$$

The constant K_e depends only upon temperature and solvent*. For the particular case in which one of the components in eq. (2.2) is the solvent, S, eq. (2.3) becomes:

*Strictly, in other than very dilute solutions, the activity coefficients, γ, for the various species A, B, and H^+, should be included in eq. (2.3) (and eq. (2.4, 2.5)) to allow for the increasing interactions between the species as the solution concentration increases. This point is discussed fully in Chapter 4

$$K_e = \frac{[B][SH^+]}{[A][S]} \quad (2.4)$$

In dilute solutions, the concentration of the solvent remains constant and may be combined with K_e, giving rise to the familiar definition of acid strength, or acid dissociation constant, $K_a (= K_e[S])$, eq. (2.5), where the symbol H^+ is used as a shorthand representation of the solvated proton, SH^+; for example, OH_3^+, $C_2H_5OH_2^+$, $(CH_3)_2SOH^+$, etc.

$$K_a = \frac{[B][H^+]}{[A]} \quad (2.5)$$

Dissociation constants so defined vary over many orders of magnitude, and it is therefore convenient to quote acid strengths as pK_a-values, eq. (2.6).

$$pK_a = -\log_{10} K_a \quad (2.6)$$

The dissociation equation to which the K_a-value refers is normally unambiguous, but in some instances, notably anilines, either of two dissociations may be relevant, e.g., $ArNH_2$:

$$ArNH_3^+ \rightleftharpoons ArNH_2 + H^+ \quad pKa(ArNH_3^+)$$

$$ArNH_2 \rightleftharpoons ArNH^- + H^+ \quad pKa(ArNH_2)$$

Anilines are predominantly encountered as bases, and it is therefore common shorthand to refer to the 'pK_a of $ArNH_2$', when what is meant is actually the 'pK_a of $ArNH_3^+$'. In cases of possible ambiguity we will identify the acid species in question.

2.2 pH and acid–base ratios†

The solution pH and the pK_a can be readily related to the relative amounts of acid, A, and base, B, present. Thus, for an acid of $pK_a < 7$ in water, a solution of the acid and its conjugate base of stoichiometric (total) concentrations $[A]_T$ and $[B]_T$, respectively, will generate the required equilibrium level of H^+ by dissociation of A, eq. (2.7).

$$A \quad \xrightleftharpoons{K_a} \quad B \quad + \quad H^+$$

concentration: $([A]_T - [H^+]) \quad ([B]_T + [H^+]) \quad [H^+]$

Hence:
$$K_a = \frac{([B]_T + [H^+])[H^+]}{([A]_T - [H^+])} \quad (2.7)$$

Thus we may write, $[H^+] = K_a([A]_T - [H^+])/([B]_T + [H^+])$ or, in terms of pH and pK_a, eq. (2.8).

$$pH = pK_a + \log_{10} \frac{[B]_T + [H^+]}{[A]_T - [H^+]} \quad \text{for low pH} \quad (2.8)$$

†At this stage we use a simple definition of pH $= -\log_{10}[H^+]$; this is strictly only applicable in very dilute solutions. A more detailed discussion of pH in aqueous and non-aqueous solvents is given in Chapter 4

Similarly, for acids with $pK_a > 7$, equilibration of the acid and its conjugate base in aqueous solution will necessarily generate an appropriate level of OH⁻, and by a similar logic to that for eq. (2.8) above, we obtain eq. (2.9); in non-aqueous solvents, SH, $[OH^-]$ is replaced by $[S^-]$.

$$pH = pKa + \log_{10} \frac{[B]_T - [OH^-]}{[A]_T + [OH^-]} \quad \text{for high pH} \quad (2.9)$$

These equations can be frequently simplified to give the more familiar eq. (2.10), provided that the concentrations of H⁺ and OH⁻ are low compared with those of A and B.

$$pH = pK_a + \log_{10} \frac{[B]_T}{[A]_T} \quad (2.10)$$

Eq. (2.10) will be valid in aqueous solution with concentrations ≥ 0.1M when combined with pH (pK_a) in the region $3 \leq pH \leq 11$. For strong acids or very weak acids in water ($pK_a \leq 3$, $pK_a \geq 11$), however, it is normally necessary to use eq. (2.8) and (2.9) when relating pH and pK_a to stoichiometric solution concentrations of A and B. In non-aqueous solvents the low tendency to self-ionize means that $[S^-]$ is almost universally very low compared with the concentrations of acids and bases present, but it remains necessary to use eq. (2.8) rather than the simpler eq. (2.10) when dealing with solutions of strong acids.

2.3 pH-dependence of species distribution

From knowledge of the pK_a of an acid it is possible in a relatively straightforward way to calculate the species distribution as a function of pH. For a simple monobasic acid, such as acetic acid, HOAc, $pK_a = 4.76$ in water at 25°C, derivation of the relationship between species distribution and pH involves a combination of equilibrium and mass-balance equations. Thus, for a given total concentration of the acid in its neutral and ionized forms, $[HOAc]_T = [HOAc] + [OAc^-]$, we may write eqs. (2.11) and (2.12), representing, respectively, the equilibrium and mass balance relationships.

$$K_a = \frac{[OAc^-][H^+]}{[HOAc]} \quad (2.11)$$

$$[HOAc]_T = [HOAc] + [OAc^-] \quad (2.12)$$

By substituting for (say) $[OAc^-] = K_a[HOAc]/[H^+]$ from eq. (2.11) into the mass balance eq. (2.12), we obtain for the fraction, α, of acetic acid in the acid form,

$$\alpha(HOAc) = \frac{[HOAc]}{[HOAc]_T} = \frac{1}{1 + K_a/[H^+]} \quad (2.13)$$

Similarly, substituting for [HOAc] from eq. (2.11) or from eq. (2.13) into eq. (2.12), gives the fraction of acetic acid in the base form, eq. (2.14).

$$\alpha(OAc^-) = \frac{[OAc^-]}{[HOAc]_T} = \frac{K_a/[H^+]}{1 + K_a/[H^+]} \quad (2.14)$$

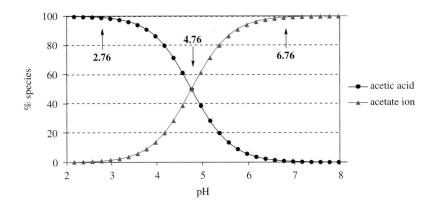

Fig. 2.1.
Species distribution for acetic acid ($pK_a = 4.76$) in water at 25°C

Eqs (2.13) and (2.14) are written in such a way that they can be readily generalized to cover polybasic acids (see Appendix 2.6).

A simple spreadsheet based on eqs. (2.13) and (2.14), allows the calculation of the species distributions as a function of solution pH; the results are shown in Fig. 2.1 for the case of acetic acid.

It can be seen that at $pH = 4.76 (= pK_a)$ the acid is 50% ionized, and it may be noted that at $pH = pK_a \pm 2$, the acid is either 99% ionized ($pH = 6.76$) or 99% in the free acid form ($pH = 2.76$), respectively.

Extension of eqs. (2.13) and (2.14) to cover polybasic systems, using the same principles of simultaneous acid–base and mass-balance equations, is presented in Appendix 2.1 [2].

Glutamic acid, for example, has four different species in equilibrium in solution, Scheme 2.1

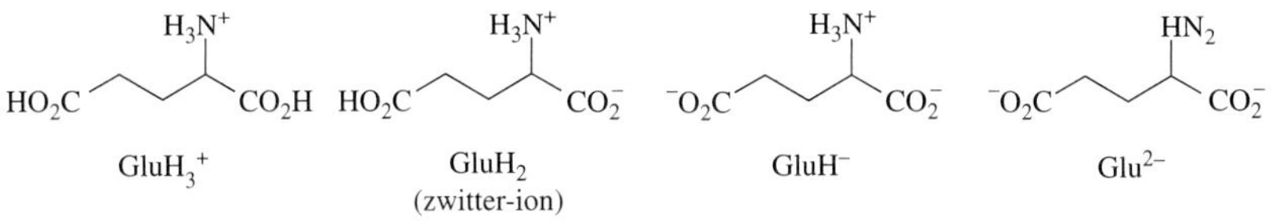

Scheme 2.1.
Ionization equilibria of glutamic acid, $GluH_2$

Application of the equations in Appendix 2.1 leads to the dependence of the proportion of the various species upon pH shown in Fig. 2.2, which also includes the pK_a-values for the successive dissociation equilibria, commencing with protonated glutamic acid, $GluH_3^+$.

There are three general features of curves such as those displayed in Fig. 2.2. The first is that the distribution curves for each successive pair of species intersect at a pH equal to the pK_a linking them; for example, $pH = 2.06$ for the cation $GluH_3^+$ and zwitter-ion $GluH_2$. The second is that the maximum proportion for each intermediate species occurs exactly midway between the pK_a-values of the neighbouring species; for example the maximum fraction of zwitter-ion occurs at $pH = 3.16 = (2.06 + 4.26)/2$. Finally, except where successive pK_a-values differ considerably ($\Delta pK_a \geq \sim 4$), quantitative formation of intermediate species *in solution* is not possible.

As an example of the practical use of such a distribution diagram, we consider optimization of the isolation of glutamic acid by crystallization. This

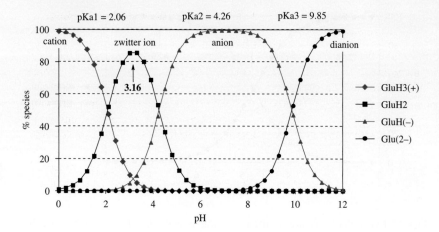

Fig. 2.2.
Species distribution for glutamic acid (GluH$_2$) in water at 25°C

*Important relationships between species distribution and reaction rates, as well as the pH-dependence of the solubility and extraction of acids and bases, are described by Atherton and Carpenter [2]

is achieved by maximization of the (neutral) zwitter-ion form in solution, which requires adjustment of the solution pH to pH = 3.16, i.e., a pH exactly midway between the pK_a-values of the neighbouring cationic and anionic forms. Furthermore, it can also be seen that there is a relatively narrow pH window in which significant proportions of the zwitter-ion exist.*

Similarly, many pharmaceutically active compounds exist in several different ionization states, and their isolation in desired cationic, zwitterionic, or anionic forms, may depend directly on the pH of the solution (and the solvent) from which they are isolated; e.g., the Aurora B Kinase Inhibitor, (I), shown as its zwitter-ion [3], has four accessible acidic sites, with pK_a-values ranging from 2.0 to 9.88.

In addition to the isolation of products by selective crystallization or extraction, knowledge of species distribution in solutions of differing pH is also often of considerable importance in elucidating reaction mechanisms and in optimizing manufacturing processes, many of which involve acid–base catalysis or coupled protolytic equilibria [2]. For example, the efficiency of an acid or base catalyst depends directly upon the fraction of the catalyst that is in the active (acid/base) form.

2.4 Acid strengths and molecular structure: dissociation constants in water

Extensive tabulations of dissociation constants in water exist, with the most comprehensive set being contained in the database ACD/pK$_a$DB^{TM}, which updates earlier compilations [4–7]. We may summarize the pK$_a$-values for common acids and bases in water schematically in Fig. 2.3. Thus, allowing for substituent effects within the various series of acids, acid strengths in water decrease in the order mineral acids > anilinium ions ∼ carboxylic acids, > phenols ∼ ammonium ions > carbon acids.

Bell [6] and Stewart [7] provide comprehensive discussions of molecular factors influencing acid strengths, and these may be conveniently summarized under three main headings.

2.4.1 The strength of the X–H bond

From a simple viewpoint it might seem that the strength of the X–H bond would be a useful indicator of acidity, but in practice the bond strength of an acid X–H, E_{XH}, is most often a very poor guide to its acidity, as illustrated by the simple hydrides:

Hydride	E_{X-H}/kJ mol^{-1}	pK$_a$ in water
CH$_4$	440	46
NH$_3$	435	35
OH$_2$	503	16

Thus, the O–H bond in water is some 70 kJ mol^{-1} stronger than the N–H bond in ammonia, and yet H$_2$O is some 20 pK units stronger as an acid than NH$_3$. Furthermore, methane has no measurable acidity in water but has the lowest bond strength. Within a closely related set of acids, however, there is a qualitative correspondence between the order of acid strengths and the H–X bond strength e.g., HF, HCl, HBr, and HI,

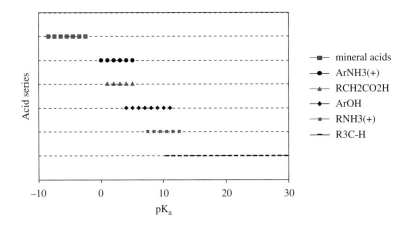

Fig. 2.3. Acid strengths in water for mineral acids, mono-substituted anilines (ArNH$_3^+$), carboxylic acids (RCH$_2$CO$_2$H), phenols (ArOH), amines (RNH$_3^+$) and carbon acids (R$_3$C − H)

Acid–Base Equilibria: Quantitative Treatment

Table 2.1 Acid strengths of inorganic oxy-acids[a]

$X(OH)_m$ very weak	pK_a	$XO(OH)_m$ weak	pK_a	$XO_2(OH)_m$ strong	pK_a	$XO_3(OH)_m$ very strong	pK_a
$Cl(OH)$	7.2	$ClO(OH)$	2.0	$ClO_2(OH)$	−1	$ClO_3(OH)$	(−10)
$Br(OH)$	8.7	$NO(OH)$	3.3	$NO_2(OH)$	−1.4	$MnO_3(OH)$	
$I(OH)$	10.0	$CO(OH)_2$	3.9	$IO_2(OH)$	0.8		
$B(OH)_3$	9.2	$SO(OH)_2$	1.9	$SO_2(OH)_2$	(−3)		
$AS(OH)_3$	9.2	$SeO(OH)_2$	2.6	$SeO_2(OH)_2$	< 0		
$Sb(OH)_3$	11.0	$TeO(OH)_2$	2.7				
$Si(OH)_4$	10.0	$PO(OH)_3$	2.1				
$Ge(OH)_4$	8.6	$AsO(OH)_3$	2.3				
$Te(OH)_6$	8.8	$IO(OH)_5$	1.6				
		$HPO(OH)_2$	1.8				
		$H_2PO(OH)$	2.0				

[a] Ref [6]

2.4.2 Charge dispersion on anions, A^-, or cations, BH^+

It is a general observation that charge dispersion stabilizes ions, and the consequence of this may be seen in a number of examples:

Inorganic oxy-acids

Striking regularities exist in the strengths of the *inorganic oxy-acids*. When expressed in the form $XO_n(OH)_m$ (for example, $H_2SO_4 \equiv SO_2(OH)_2$), the acids fall into four groups, listed in Table 2.1, depending upon the value n, but essentially independent of the number of OH groups, m, or the nature of the central atom, X.

Explanations for the regularities in Table 2.1 relate to the number of equivalent oxygen atoms in the acid anion, which increases from one for the very weak acids, $X(OH)_n$, to four for the very strong acids, $XO_3(OH)$. This in turn determines the number of oxygen atoms over which the negative charge may be distributed, e.g., $SO_3(OH)^- (HSO_4^-)$ vs $SO_2(OH)^- (HSO_3^-)$:

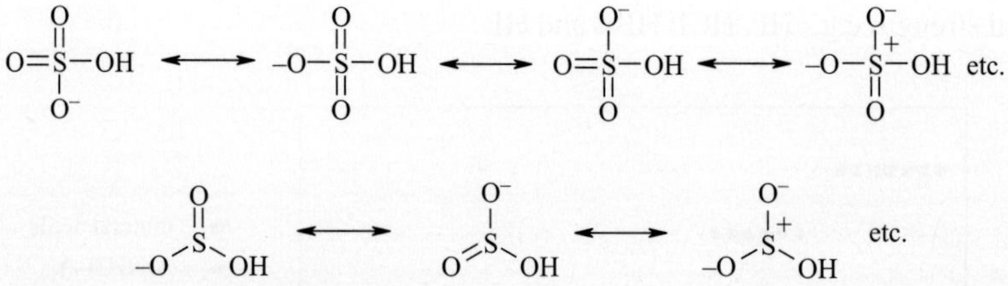

Carboxylic acids versus alcohols

Carboxylic acids are much stronger than simple alcohols, and this can be related to the structure of the carboxylate anion, which allows distribution of the charge across two oxygen atoms:

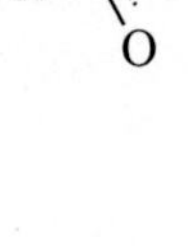

Amines, amidines and guanidines

The dissociation constants of the conjugate acids of amines vary significantly with the ability to stabilize the cationic charge by dispersion:

amine $RNH_2 \longrightarrow RNH_3^+$ (moderate base, pK_a c 10.5)

amidine $R-C(NH_2)=NH \longrightarrow R-C(NH_2)_2^+$ (stronger, pK_a c 12)

guanidine $HN=C(NH_2)_2 \longrightarrow [H_2N-C(NH_2)_2]^+$ (very strong, pK_a c 14)

2.4.3 The nature of X in R–X–H

The nature of atom, X, in R–X–H has a significant effect on the acidity. In general, for a related series of compounds, changing from a –CH to an –NH acid leads to a decrease of $\sim$ 3pK units and from an –NH to an –OH acid, a further decrease of $\sim$ 10 units; e.g., compare **CH$_3$NO$_2$**($pK_a = 10.2$), **NH$_2$NO$_2$** ($pK_a = 6.5$), and **HONO$_2$**($pK_a = -1.3$). Similarly, thio-acids are stronger than their oxygen analogues; ArSH($pK_a = 6.6$) *versus* ArOH($pK_a = 10.0$).

2.5 Carbon acids

Special mention may be made of carbon acids, i.e., acids in which the proton is attached to carbon, because of their importance in a wide range of synthetic procedures. The paraffin hydrocarbons have no detectable acidic properties, and among aromatic hydrocarbons only those which contain extensive π-electron systems exhibit significant acidity; for example fluorene, $pK_a = 21$, often used as an indicator in non-aqueous solvents [8].

fluorene

The negative charge on the conjugate base anion can be distributed over a number of atoms, which can be formally represented through resonance structures:

etc.

Higher acidities occur for carbon acids in which ionization is accompanied by a structural rearrangement, allowing transfer of the negative charge on the conjugate base to an oxygen atom, which has a much higher electron affinity than carbon. Examples include esters, ketones, and nitroalkanes:

Esters [9]

$$\text{HC}-\text{C}(=O)(OR) \rightleftharpoons \text{C}(=C)(O^-)(OR) + H^+ \quad pK_a = 25.6 \text{ (ethyl acetate)}$$

The related pK_a of the methyl group of acetic acid in water has also been measured as $pK_a = 22.7$ [10].

Ketones [11]

$$\text{HC}-\text{C}(=O) \rightleftharpoons \text{C}=\text{C}(O^-) + H^+ \quad pK_a = 19.3 \text{ (acetone)}$$

Nitroalkanes [12]

$$\text{HC}-\text{N}^+(=O)(O^-) \rightleftharpoons \text{C}=\text{N}^+(O^-)(O^-) + H^+ \quad pKa = 10.2 \text{ (nitromethane)}$$

Table 2.2 lists the dissociation constants in water for important examples.

The acidity of hydrocarbons is also increased sharply by inclusion of strongly electron-withdrawing substituents, such as –CN and –SO$_2$R. Although resonance structures in which the charge is delocalized to nitrogen or oxygen can often be written for many of these species, most evidence suggests that decreases in pK_a arise predominantly from polar rather than mesomeric effects [13]. Table 2.3 lists dissociation constants for representative carbon acids in which the C-H bond is activated by a polar substituent, X.

The high pK_a-values of these important classes of carbon acids serve to highlight the difficulty of generating significant quantities of their anions in water. It follows from eq. (2.14), which relates the extent of ionization to pH, that, for example, acetone, pK_a 19.3 (Table 2.2), will be ionized to an extent of only 1 part in 10^5 at pH 14, i.e., only 1 part in 10^5 will exist in the form of the reactive enolate ion. This proportion falls to 1 part in 10^{11} for the anion of acetonitrile, and similarly for the other substrates included in

Table 2.2 pK_a-values of ketones and related substances, CH$_3$ X, in water at 25°C[a]

Group X	pK_a	Group X	pK_a
CO_2^-	33.5	COSEt	21.0
$CONH_2$	28.4	$COCH_3$	19.3
CO_2Et	25.6	COPh	18.2
CO_2H	22.7[b]	NO_2	10.2

[a] Rezende, M.C. *Tetrahedron*, 2001, *57*, 5923; [b] Ref [10]

Table 2.3 pK_a-values of carbon acids, CH_3X in water at 25°C[a]

Group X	SOCH$_3$	$(CH_3)_3P^+$	$(CH_3)_2S^+$	CN	SO$_2$CH$_3$
pK_a	33	29.4[b]	28.5[b]	28.9[c]	23

[a] Rezende, M.C. *Tetrahedron*, 2001, *57*, 5923; [b] Rios, A; O'Donoghue, A.C.; Amyes, T.L.; Richard, J.P. *Can. J. Chem.*, 2005, *83*, 1536; [c] Richard, J.P.; Williams, G.; Gao, J. *J. Am. Chem. Soc.*, 1999, *121*, 715

Tables 2.2 and 2.3. Thus, substrates with pK_a-values substantially in excess of 14 can only be ionized to a very small extent in water.

2.6 Acid–base equilibria

Individual pairs of acids and their conjugate bases may be combined as in eq. (2.2) to generate an equilibrium mixture of the different species. It follows simply from eq. (2.3) that the equilibrium constant for such a mixture, K_e, is given by the ratio of the two individual acidity constants, i.e., $K_e = K_{a1}/K_{a2}$, and hence for any pair of acids and bases:

$$\log K_e = pK_{a2} - pK_{a1} \qquad (2.15)$$

For a typical carboxylic acid, acetic acid, $pK_a = 4.76$ at 25°C, and amine, triethylamine, pK_a (triethylammonium) $= 10.75$, for example, the equilibrium and equilibrium constant is given by eqs. (2.16) and (2.17).

$$CH_3CO_2H + Et_3N \underset{H_2O}{\overset{K_e}{\rightleftharpoons}} CH_3CO_2^- + Et_3NH^+ \qquad (2.16)$$

$$K_e = 10^{(10.75-4.76)} = 10^{5.99} = 1.0 \times 10^6 \qquad (2.17)$$

In this case, the equilibrium lies overwhelmingly to the right. We will see in subsequent chapters, however, that the value of this and other such equilibrium constants may depend very strongly upon the solvent.

Appendix 2.1 Species distribution in acid-base systems [2]

The acid H_2A has two coupled acid–base equilibria, represented by eq. (A2.1).

$$H_2A \overset{K_{a1}}{\rightleftharpoons} HA^- \overset{K_{a2}}{\rightleftharpoons} A^{2-} \qquad (A2.1)$$

The relevant equilibria and mass-balance equations are given by eqs. (A2.2)–(A2.4).

$$K_{a1} = \frac{[H^+][HA^-]}{[H_2A]} \quad (A2.2)$$

$$K_{a2} = \frac{[H^+][A^{2-}]}{[HA^-]} \quad (A2.3)$$

$$[HA]_t = [H_2A] + [HA^-] + [A^{2-}] \quad (A2.4)$$

Successive substituting for the various species from eqs. (A2.2) and (A2.3) in terms of $[H^+]$ into eq. (A2.4) gives eqs. (A2.5)–(A2.7) for the fraction of the various species, H_2A, HA^-, and A^{2-} as a function of the hydrogen ion concentration (pH) and the pK_a-values.*

*Thus, for example, substituting $[HA^-] = (K_{a1}/[H^+])[H_2A]$ and $[A^{2-}] = (K_{a2}/[H^+])[HA^-] = (K_{a1}K_{a2}/[H^+]^2)[H_2A]$ into eq. (A2.4) gives eq. (A2.5), etc.

$$\frac{[H_2A]}{[H_2A]_T} = \frac{1}{1 + K_{a1}/[H^+] + K_{a1}K_{a2}/[H^+]^2} \quad (A2.5)$$

$$\frac{[HA^-]}{[H_2A]_T} = \frac{K_{a1}/[H^+]}{1 + K_{a1}/[H^+] + K_{a1}K_{a2}/[H^+]^2} \quad (A2.6)$$

$$\frac{[A^{2-}]}{[H_2A]_T} = \frac{K_{a1}K_{a2}/[H^+]^2}{1 + K_{a1}/[H^+] + K_{a1}K_{a2}/[H^+]^2} \quad (A2.7)$$

Extension of eqs. (A2.5)–(A2.7) to cover the general case of acid H_nA is obvious.

Provided the various pK_a are known, eqs. (A2.5)–(A2.7) and their analogues for other acids, H_nA, can be used in conjunction with a simple spreadsheet to calculate the pH-dependence of the species distribution.

References

[1] Bell, R. P. 'Acids and Bases: Their Quantitative Behaviour', Methuen, 1971
[2] Atherton, J. H., Carpenter, K. J. 'Process Development: Physicochemical Concepts', OUP, 1999
[3] Wilkinson, R. W., Odedra, R., Heaton, S. P., Wedge, S. R., Keen, N. J., Crafter, C., Foster, J. R., Brady, M. C., Bigley, A., Brown, E., Byth, K. F., Barrass, N. C., Mundt, K. E., Foote, K. M., Heron, N. M., Jung, F. H., Mortlock, A. A., Boyle, F. T., Green, S. *Clinical Cancer Res.*, 2007, *13*, 3682
[4] Christensen, J. J., Hansen, L. D., Izatt, R. M. 'Handbook of Proton Ionization Heats and Related Thermodynamic Quantities', Wiley-Interscience, 1976
[5] Perrin, D. D, Dempsey, B., Serjeant, E. P. 'pK_a Predictions for Organic Acids and Bases', Chapman & Hall, 1981
[6] Bell, R. P. 'The Proton in Chemistry', Chapman and Hall, 1973
[7] Stewart, R. 'The Proton: Applications to Organic Chemistry', Academic Press, NY, 1985
[8] Bordwell, F. G., Branca, J. C., Hughes, D. L., Olmstead, W. N. *J. Org. Chem.* 1980, *45*, 3305
[9] Amyes, T. L., Richard, J. P. *J. Am. Chem. Soc.*, 1996, *118*, 3129
[10] Grabowski, J. J. *Chem. Commun.*, 1997, 255
[11] Kresge, A. J. *Acc. Chem. Res.*, 1990, *23*, 43
[12] Taft, R. W., Bordwell, F. G. *Acc. Chem. Res.*, 1988, *21*, 463
[13] Goumont, R., Magnier, E., Kizilian, E., Terrier, F. *J. Org. Chem.*, 2003, *68*, 6566

Solvation and Acid–Base Strength

3

Neutral acids, HA, such as carboxylic acids and phenols, eq. (3.1), are often very much weaker in non-aqueous solvents than in water or, equivalently, their anionic conjugate bases are very much stronger. On the other hand, the acidity of protonated amines and related nitrogen bases, eq. (3.2), is much less dependent on solvent.

$$\text{PhOH} \rightleftharpoons \text{H}^+ + \text{PhO}^- \quad (3.1)$$

$$\text{R}_3\text{NH}^+ \rightleftharpoons \text{H}^+ + \text{R}_3\text{N} \quad (3.2)$$

A brief inspection of eqs. (3.1) and (3.2) suggests a possible underlying reason for their different sensitivities to solvent; namely, that the dissociation of PhOH results in the generation of two charged species, whereas there is no change in charge on dissociation of R_3NH^+.

In general, we expect charged species to be much more sensitive to solvent changes than neutral species, and indeed we will see that the dominant influence of a solvent lies in its ability to stabilize charge.

In this chapter we consider the influence of solvent on the (free) energy of electrolytes and non-electrolytes and the relationship between these changes in free energy and the dissociation constants in the solvents; e.g., in eq. (3.1) the effect of solvent on the free energies of PhOH, H^+ and PhO^-. The solvation energies provide a basis for the understanding of the solvent-dependence of dissociation constants and they can also be used to enable reliable estimates of pK_a in the absence of directly measured values. For convenience, we use water as the reference solvent from which the species are transferred, but the change in free energy between any other pair of solvents can be readily obtained by difference.

3.1 Solvation and acid dissociation constants: free energies of transfer

The difference in free energy of a species in two solvents is termed the free energy of transfer, ΔG_{tr}. The relationship between ΔG_{tr} for the various acid–base species and the change in pK_a with solvent—for example, between water and solvent S—is best illustrated using a Born–Harber cycle, as in Scheme 3.1.

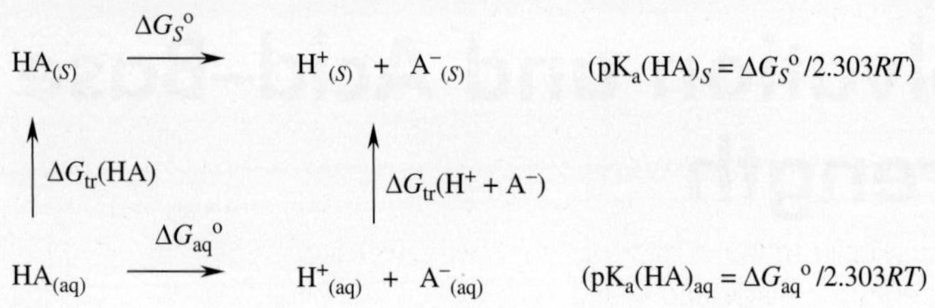

Scheme 3.1.
Born–Haber cycle for the dissociation of acid HA

$$HA_{(S)} \xrightarrow{\Delta G_S^o} H^+_{(S)} + A^-_{(S)} \qquad (pK_a(HA)_S = \Delta G_S^o/2.303RT)$$

$$\uparrow \Delta G_{tr}(HA) \qquad \uparrow \Delta G_{tr}(H^+ + A^-)$$

$$HA_{(aq)} \xrightarrow{\Delta G_{aq}^o} H^+_{(aq)} + A^-_{(aq)} \qquad (pK_a(HA)_{aq} = \Delta G_{aq}^o/2.303RT)$$

In Scheme 3.1, $\Delta G_{tr}(HA)$ represents the change in free energy of the undissociated acid on transfer from water to solvent S, and similarly for $\Delta G_{tr}(H^+ + A^-) = \Delta G_{tr}(H^+) + \Delta G_{tr}(A^-)$. ΔG_S^o and ΔG_{aq}^o are the free energies of dissociation of HA in solvent S and water, respectively. It follows from the standard relationship between equilibrium constant, K, and the corresponding free energy change, ΔG, i.e., $\Delta G = -RT\ln K$, that the pK_a and the free energy of dissociation, ΔG^o, of HA are related by eq. (3.3).*

*Note that the factor of 2.303 in Scheme 3.1, eq. (3.3) and subsequent equations arises from the change from natural logarithms ($\ln K_a = -\Delta G^o/RT$) to logarithms to the base 10: $pK_a = -\log K_a = -\ln K_a/2.303$

$$pK_a = -\log K_a = -(\ln K_a)/2.303 = \Delta G^o/2.303RT \qquad (3.3)$$

The Born–Haber cycle in Scheme 3.1 shows furthermore that

$$\Delta G_S^o = \Delta G_{aq}^o + \Delta G_{tr}(H^+) + \Delta G_{tr}(A^-) - \Delta G_{tr}(HA)$$

Thus, by combining the expression for ΔG_S^o with eq. (3.3), it follows that the change in pK_a of acid HA on transfer between water and solvent S, ΔpK_a, is given by eq. (3.4), and similarly eq. (3.5) for acid BH^+.

$$pK_a(HA)_S - pK_a(HA)_{aq} = (\Delta G_s^o - \Delta G_{aq}^o)/2.303RT$$
$$= \{\Delta G_{tr}(H^+) + \Delta G_{tr}(A^-)$$
$$- \Delta G_{tr}(HA)\}/2.303RT \qquad (3.4)$$

$$pK_a(BH^+)_S - pK_a(BH^+)_{aq} = (\Delta G_s^o - \Delta G_{aq}^o)/2.303RT$$
$$= \{\Delta G_{tr}(H^+) + \Delta G_{tr}(B)$$
$$- \Delta G_{tr}(BH^+)\}/2.303RT \qquad (3.5)$$

Eqs (3.4) and (3.5) show, therefore, that we should be able to understand and, in principle, predict changes in pK_a-values with solvent from the influence of solvent on the free energies of the various participants in the acid–base equilibria. We may note that the changes in solvation energies of ions on solvent transfer are normally considerably higher than those of neutral species, especially when differences between two non-aqueous solvents are considered (see Section 3.4), and hence are expected to dominate the corresponding changes in pK_a-values.

In numerical terms, it follows from the relationship between pK_a and ΔG, eq. (3.3) that at 25°C:†

†$2.303\,RT = 5710\,J\,mol^{-1}$ at 25°C

$$1\,pK\,unit \equiv 5.7\,kJ\,mol^{-1} \qquad (3.6)$$

Thus, a free energy change of 5.7 kJ mol^{-1} equates to a change of one unit in pK_a.

3.1.1 Solvent-transfer activity coefficients

An equally valid and widely used alternative means of reporting the changes in free energy occurring when a species Y is transferred from a reference solvent O (in our case water) to solvent S, is via *solvent-transfer activity coefficients* ($^O\gamma^S$), also called medium effects or degenerate activity coefficients [1, 2]. They are related to the corresponding free energies of transfer by eq. (3.7), for transfer between water and solvent S.‡

$$\Delta G_{tr}(Y) = RT \ln {}^{aq}\gamma^S(Y) \qquad (3.7)$$

In the case of electrolyte MX, the analogous equation is eq. (3.8).

$$\Delta G_{tr}(MX) = \Delta G_{tr}(M^+) + \Delta G_{tr}(X^-)$$
$$= RT \ln {}^{aq}\gamma^S(M^+){}^{aq}\gamma^S(X^-) \qquad (3.8)$$

‡Solvent-transfer activity coefficients reflect changes in *solute–solvent interactions* in different solvents, and hence may be very large. More familiar is the use of activity coefficients to quantify the much smaller, concentration-dependent *solute–solute interactions* in a given solvent, such as electrostatic attractions between ions of opposite charge (Chapter 4)

Free energies of transfer and transfer activity coefficients may, therefore, be used interchangeably to represent changes in solvation energy.

The change in K_a of acid HA with solvent can also be readily expressed in terms of solvent-transfer activity coefficients, as in eqs. (3.9) and (3.10).

$$K_a(HA)_{aq}/K_a(HA)_S = {}^{aq}\gamma^S(H^+){}^{aq}\gamma^S(A^-)/{}^{aq}\gamma^S(HA) \qquad (3.9)$$

$$pK_a(HA)_S - pK_a(HA)_{aq} = \log{}^{aq}\gamma^S(H^+) + \log{}^{aq}\gamma^S(A^-)$$
$$- \log{}^{aq}\gamma^S(HA) \qquad (3.10)$$

There is, therefore, a direct and obvious relationship between solvent-transfer activity coefficients ($\log {}^{aq}\gamma^S$) and changes in pK_a-values.

The main advantage of using $\log {}^{aq}\gamma^S$ values, as in eq. (3.10), is the numerically simple relationship between changes in pK_a and $\log {}^{aq}\gamma^S$: a change of 1 unit in $\log {}^{aq}\gamma^S$ for any of the species involved in the equilibria corresponds directly to a change of 1 unit in pK_a.

Most chemists, however, are more comfortable with the notion of free energies than with activity coefficients, and we will, therefore, normally use free energies of transfer in preference to solvent-transfer activity coefficients to express changes in solvation energies and hence equilibrium constants. Occasionally, it will be convenient also to include $\log {}^{aq}\gamma^S$ values because of their direct equivalence to changes in pK_a.

3.2 Determination of free energies of transfer

3.2.1 Non-electrolytes

(i) *Solubility measurements*. The most widely used method for determining the change in free energy of a species Y between solvents is based on measurement of the solubility of Y in the different solvents.

The free energy of solution of substrate Y in a given solvent, $\Delta G_s(Y)$, is related to its solubility, S_o, in that solvent by eq. (3.11).

$$\Delta G_s(Y) = -RT \ln S_o \qquad (3.11)$$

Thus, the difference in the solubility of Y in two different solvents can be used to determine the change in energy, $\Delta G_{tr}(Y)$, on transfer between the solvents.

The method is illustrated for benzoic acid, via the simple Born–Haber cycle shown in Scheme 3.2, in which ΔG_s^{aq} and ΔG_s^S represent the free energies of solution of benzoic acid in water and solvent S, respectively.

$$\begin{array}{ccc} ArCO_2H_{(aq)} & \xrightarrow{\Delta G_{tr}} & ArCO_2H_{(S)} \\ {}_{\Delta G_s^{aq}} \nwarrow & & \nearrow {}_{\Delta G_s^S} \\ & ArCO_2H_{(c)} & \end{array}$$

Scheme 3.2. Born–Haber cycle for the solubility of benzoic acid

*Provided the acid does not form a solvate in the solid state, i.e., the solid state is unaffected by the change in solvent

†The equivalent expression for the solvent transfer activity coefficient is $\log{}^{aq}\gamma^S = -\log(S_o^S/S_o^{aq})$

It is apparent from Scheme 3.2 that changes in solubility are a direct consequence of the difference in the free energy of benzoic acid in the two solvents.* Thus, the free energy change of benzoic acid on transfer between water and solvent S, ΔG_{tr} (benzoic acid), and its solubility in the two solvents are related by eq. (3.12).†

$$\Delta G_{tr} \text{ (benzoic acid)} = \Delta G_s^S - \Delta G_s^{aq} = -RT \ln (S_o^S/S_o^{aq}) \quad (3.12)$$

Table 3.1 lists the measured solubility of benzoic acid in water and some common solvents [3], together with the derived free energies of solution and transfer and the equivalent solvent-transfer activity coefficients.

The free energy of benzoic acid in the non-aqueous solvents is thus ~ 10kJ mol^{-1} lower than in water, reflecting the more effective solvation of benzoic acid. This decrease in free energy, which reduces its tendency to dissociate, contributes an increase of ~ 2 units to the pK_a of benzoic acid on transfer from water (eqs. (3.4), (3.10)).

(ii) *Vapour pressure measurements.* A second commonly used method for the determination of free energies of transfer of non-electrolytes is based on the solvent-dependence of vapour pressure. The variation in vapour pressure of a volatile substrate with solvent is controlled directly by the difference in solvation in an analogous manner to that of its solubility. For a given solvent, the vapour pressure of substrate Y, P_Y, is proportional to its concentration, as expressed by Henry's Law, eq. (3.13).

$$P_Y \propto [Y]$$
$$= H_Y[Y] \quad (3.13)$$

Table 3.1 Solubility and free energies of transfer of benzoic acid at 25°C[a]

Benzoic acid	H$_2$O	MeOH[b]	DMF[b]	MeCN[b]
S_o/M	0.0278	3.16	5.3	0.85
ΔG_s/kJ mol^{-1}	8.87	−2.85	−4.13	0.40
ΔG_{tr}/kJ mol^{-1} [c]	0	−11.7	−13.0	−8.5
$\log{}^{aq}\gamma^{Sd}$	0	−2.0	−2.3	−1.5

[a] Ref [3]; [b] Abbreviations as in Table 1.1; [c] Free energy of transfer from water to solvent; [d] ${}^{aq}\gamma^S = S_o^S/S_o^{aq}$ is the solvent-transfer activity coefficient (Section 3.1.1)

In eq. (3.13) the proportionality constant between pressure and concentration, H_Y, is known as the *Henry's Law constant*. It is determined experimentally by measuring the vapour pressure of Y over solutions of known concentration of Y.

The free energy of transfer of Y between water and solvent, S, is related to the change in Henry's Law constant by eq. (3.14).

$$\Delta G_{tr}(Y) = RT \ln(H_Y^S/H_Y^{aq}) \qquad (3.14)$$

Thus, measurement of Henry's Law constant, or, equivalently, the vapour pressure of Y at constant concentration in different solvents, can be used to determine ΔG_{tr}-values.

(iii) *Distribution coefficients*. Finally, ΔG_{tr}-values for non-electrolytes may also be determined from measurement of distribution coefficients.[‡] The distribution coefficient for substrate Y between water and any non-miscible solvent, S, $D = [Y]^S/[Y]^{aq}$ is linked to the corresponding free energy of transfer of Y by eq. (3.15).

$$\Delta G_{tr}(Y) = -RT \ln(D) \qquad (3.15)$$

[‡]The most familiar example of the use of distribution coefficients is that of log P values, where P is the partition coefficient of a substrate between water and the immiscible solvent 1-octanol. They are widely used to provide a measure of the bioavailability of agrochemical and pharmaceutical products

The method is of limited use as it is restricted to pairs of immiscible solvents—typically water and non-polar organic solvents. It can be used indirectly for water-miscible solvents, such as methanol, dimethylformamide, acetonitrile, etc., by comparing the separate partitioning of Y between the two solvents and a third, immiscible solvent, such as decalin or cyclohexane.

In each of the above methods, the free energy of transfer is based on equilibration of the solute between the solvent in question and a common (invariant) reference phase: the solid state (solubility measurements); the vapour phase (Henry's Law); or an immiscible solvent (partition coefficients).

3.2.2 Electrolytes

(i) *Solubility measurements*. For a typical salt, such as sodium chloride, the solubility in different solvents, $S_o = [NaCl]_{sat} = [Na^+]_{sat} = [Cl^-]_{sat}$, is related to the corresponding free energy of transfer of the constituent ions in an analogous manner to that described above for non-electrolytes. The solubility product, $K_{sp} = [Na^+][Cl^-] = S_o^2$, is related to the free energy of solution, $\Delta G_s(NaCl)$, by eq. (3.16).

$$\Delta G_s(NaCl) = -RT \ln K_{sp} = -2RT \ln S_o \qquad (3.16)$$

The change in free energy of NaCl with solvent is illustrated in Scheme 3.3, which shows a Born–Haber cycle for the solution of NaCl in water and solvent.[§]

[§]An assumption inherent in Scheme 3.3 is that the salt does not form a solvate with any of the solvents involved; this can be tested by analysis of the solid in equilibrium with solvent

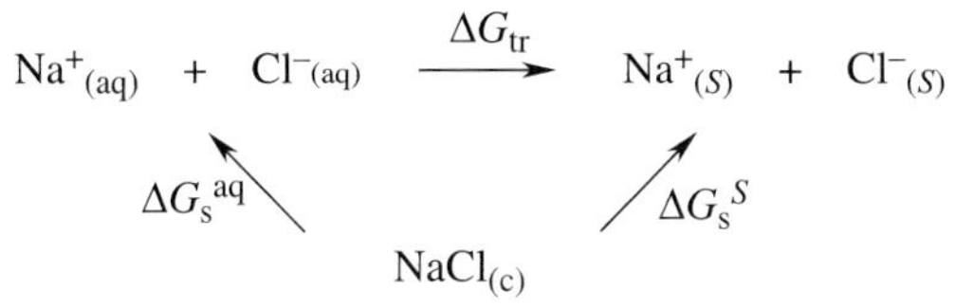

$$\begin{array}{ccc} Na^+_{(aq)} + Cl^-_{(aq)} & \xrightarrow{\Delta G_{tr}} & Na^+_{(S)} + Cl^-_{(S)} \\ {}_{\Delta G_s^{aq}}\nwarrow & & \nearrow_{\Delta G_s^S} \\ & NaCl_{(c)} & \end{array}$$

Scheme 3.3.
Born–Haber cycle for the solubility of sodium chloride

Table 3.2 Solubility and free energies of transfer of NaCl at 25°C[a]

NaCl	H_2O	MeOH[b]	DMF[b]	MeCN[b]
S_o/M	7.0	6.3×10^{-2}	3.2×10^{-3}	7.2×10^{-5}
K_{sp}/M^2	49	3.9×10^{-3}	1.0×10^{-5}	5.2×10^{-9}
ΔG_s/kJ mol^{-1}	-9.00[c]	13.72	28.49	47.28
ΔG_{tr}/kJ mol^{-1} [d]	0	22.7	37.5	56.3
$\log\,^{aq}\gamma^{S}$[e]	0	3.9	6.6	9.9

[a] Ref [2]; [b] Abbreviations as in Table 1.1; [c] Free energy of solution at infinite dilution: Rossini, F.D., et al., U.S. Nat. Bureau of Standards, Circular 500 and supp. Notes 270–1 to 270–3; [d] Free energy of transfer from water to solvent; [e] Solvent-transfer activity coefficient

*Equivalently,
$\log\,^{aq}\gamma^{S} = -\log(S_o^S/S_o^{aq})$

It follows from Scheme 3.3 and eq. (3.16) that the change in free energy is given by eq. (3.17).*

$$\Delta G_{tr}(Na^+) + \Delta G_{tr}(Cl^-) = \Delta G_s^S - \Delta G_s^{aq} = -2RT\,\ln(S_o^S/S_o^{aq}) \quad (3.17)$$

Table 3.2 lists the measured solubility of sodium chloride in several solvents, together with the derived free energies of solution and transfer.

These effects are normally very large compared with those for non-electrolytes: the transfer of NaCl from water to acetonitrile, for example, is accompanied by a decrease in solubility product of some *10 orders of magnitude*. We can anticipate similar, if not larger, effects when we come to examine acid–base dissociation and equilibria in different solvents.

(ii) *EMF measurements*. The solvent dependence of the EMF of appropriate electrochemical cells provides another important source of free energy data for electrolytes. Consider, for example, the following cell and its corresponding half-cell reactions, eq. (3.18):

$$Pt, Cl_{2(g)} \mid HCl_{(aq)} \mid H_{2(g)}, Pt$$

$$H^+_{(aq)} + e \rightarrow \tfrac{1}{2}H_{2(g)}$$
$$\tfrac{1}{2}Cl_{2(g)} + e \rightarrow Cl^-_{(aq)} \quad (3.18)$$

The overall cell reaction is given by eq. (3.19).

$$H^+_{(aq)} + Cl^-_{(aq)} \rightarrow \tfrac{1}{2}H_{2(g)} + \tfrac{1}{2}Cl_{2(g)} \quad (3.19)$$

The free energy change, ΔG_{aq}, for the cell reaction is related to the EMF of the cell, E_{aq}, by eq. (3.20), in which F is the Faraday constant.

$$\Delta G_{aq} = -F.E_{aq} \quad (3.20)$$

The essential feature of cell (3.18) is that *changes in EMF arising from a change in solvent are due entirely to changes in the free energies of the ions*, H^+ and Cl^-. This is because the free energies of the (gaseous) products of the cell reaction, $H_{2(g)}$ and $Cl_{2(g)}$ (eq. (3.19)), are independent of solvent. Thus, if the corresponding EMF in solvent S, E_S, is subtracted from that of the aqueous cell, eq. (3.20), the net process equates to the transfer of $H^+ + Cl^-$ from water to solvent S, eq. (3.21), and free energy change is given by eq. (3.22).

Free energies of ion solvation

Table 3.3 Electrochemical cells for determining free energies of transfer

Cell[a]	Cell reaction
Ag, AgCl$_{(c)}$ $\|$HCl$_{(S)}\|$H$_{2(g)}$, Pt[b]	H$^+_{(S)}$ + Cl$^-_{(S)}$ + Ag → ½H$_{2(g)}$ + AgCl$_{(c)}$
Ag, AgBzO$_{(c)}$ $\|$BzO$^-_{(S)}\|$H$^+_{(S)}\|$H$_{2(g)}$, Pt[c]	H$^+_{(S)}$ + BzO$^-_{(S)}$ + Ag → ½H$_{2(g)}$ + AgBzO$_{(c)}$
Pt, Na(Hg)$\|$Na$^+_{(S)}\|\|$Ag$^+_{(S)}\|$Ag	Ag$^+_{(S)}$ + Na(Hg) → Na$^+_{(S)}$ + Ag

[a] BzO$^-$ = benzoateion. [b] Cell saturated with respect to AgCl. [c] Left hand half-cell saturated with respect to silver benzoate

$$H^+_{(aq)} + Cl^-_{(aq)} \longrightarrow H^+_{(S)} + Cl^-_{(S)} \qquad (3.21)$$

$$\Delta G_{aq} - \Delta G_S = \Delta G_{tr}(H^+) + \Delta G_{tr}(Cl^-) = -F.(E_{aq} - E_S) \qquad (3.22)$$

Table 3.3 lists other examples of electrochemical cells that can be used to determine free energies of transfer of electrolytes, together with the overall cell reactions.

In each case, the only components whose free energies change with solvent are the ions; the remaining cell components are either solids or gases and hence their free energies are solvent-independent. Thus, the cells listed in Table 3.3, can be used to determine ΔG_{tr} values for (H$^+$ + Cl$^-$), (H$^+$ + BzO$^-$), and (Ag$^+$ − Na$^+$), respectively.

(iii) *Distribution coefficients.* Finally, in a small number of cases in which the non-aqueous solvent is immiscible with water, e.g., nitrobenzene, free energies of transfer may be determined directly from the partitioning of salts between the two solvents, analogous to the partition method noted above for non-electrolytes (eq. (3.15)).

It is very important to note that all measurements on electrolyte solutions *involve electrically neutral combinations of ions* and hence can only give data pertaining to such combinations. These are either whole electrolytes, as in the case of the solubility of NaCl, or differences between ions of like charge, as in the final cell in Table 3.3, which measures the *difference* between Na$^+$ and Ag$^+$ on solvent transfer.[†]

[†]An exception is Volta-potential measurements (equivalent to electron work functions) which measure the total energy change in removing an ion from solution to the gas-phase: Farrell, J.R., McTigue, P.T. *J. Electroanal. Chem.,* 1982, *139*, 37

3.3 Free energies of ion solvation

3.3.1 Hydration of ions

In the first instance, we consider briefly the absolute free energies of solvation of ions, i.e., the transfer of ions from the gas-phase to solution, eq. (3.23) for cations and similarly for anions.

$$M^{n+}_{(g)} + S_{(l)} \longrightarrow M^{n+}_{(S)} \qquad (3.23)$$

The energies involved are enormous, comparable to the lattice energies of ionic crystals, as illustrated by the hydration energies of ions (eq. (3.23), S = H$_2$O) listed in Table 3.4 [4]; values range from hundreds to thousands of kJ mol^{-1}.

The absolute values are dominated by the very large electrostatic energies of the ions in the gas relative to the solution phase. Nevertheless, even in the

Table 3.4 Free energies of hydration of ions at 25°C[a]

Ion	$r/\text{Å}^b$	$-\Delta G°_h$/kJ mol^{-1}	Ion	$r/\text{Å}^b$	$-\Delta G°_h$/kJ mol^{-1}
H$^+$		1089	Al^{3+}	0.52	4615
Li$^+$	0.60	511	Sc^{3+}	0.75	3885
Na$^+$	0.95	411	Y^{3+}	0.93	3541
K$^+$	1.33	337			
Rb$^+$	1.48	316	F$^-$	1.36	436
Cs$^+$	1.69	284	Cl$^-$	1.81	311
Ag$^+$	1.26	473	Br$^-$	1.95	285
Be^{2+}	0.31	2442	I$^-$	2.16	247
Mg^{2+}	0.65	1906	OH$^-$	1.40	403
Ca^{2+}	0.99	1593	CN$^-$	1.90	310
Sr^{2+}	1.13	1447	NO$_3^-$	1.89	270
Ba^{2+}	1.35	1318	ClO$_4^-$	2.36	178

[a] Ref [4]; [b] Pauling crystal radii for monatomic ions; thermochemical radii for polyatomic ions

gas phase *the most significant interactions between ions and solvent molecules are those occurring within the first solvation sphere*. This is shown by mass-spectrometric measurements on the association of ions with solvent molecules in the gas phase [5]. Thus, the equilibrium constants for the successive addition of water molecules to ions, eq. (3.24), show that beyond the first 5–6 water molecules, i.e., the first solvation shell, the decrease in free energy per additional water molecule is almost independent of the cation.

$$[M(H_2O)_{n-1}]^+_{(g)} + H_2O_{(g)} \rightleftharpoons [M(H_2O)_n]^+_{(g)} \tag{3.24}$$

Furthermore, the energy gain from addition of water molecules to the second-shell is very close to the free energy of condensing a water molecule from the gas phase to liquid water ($\Delta G \sim -9$ kJ mol^{-1}). Similar results were obtained for other common solvent molecules, such as methanol and acetonitrile.

3.3.2 Solvation in pure solvents

Changes in ion-solvent interactions on transfer of electrolytes between solvents are much smaller than the absolute solvation energies and differences in electrostatic energies play a much reduced role. They are nevertheless sufficiently large to cause dramatic changes in chemical reactions and equilibria involving ions. As expected, *the changes result primarily from differences in specific interactions of the ions with the immediate-neighbour solvent molecules*; for example, ion-dipole, Lewis acid–base and H-bonding interactions (Section 1.2.1).

Solubility data for simple electrolytes in polar solvents have been summarized by Johnsson and Persson [6], and from these and complementary measurements on appropriate electrochemical cells it is possible to derive the free energies of transfer of a wide range of electrolytes from water to a variety of solvents [2, 7–11]. Although experimental measurements of the free energies refer to whole electrolytes (or differences between ions of like charge), it is convenient to report them in terms of individual ions. The use of anionic and cationic values facilitates a comparison within groups of anions or cations, and the individual values can also be recombined as appropriate to give values for a much wider set of electrolytes.

The division into ionic values could be achieved by arbitrarily assigning a value for, say, the proton and then reporting all other values relative to this. More satisfying, however, is the most widely used convention used for the reporting of individual values, whereby the free energies of transfer of Ph_4As^+ and Ph_4B^- are equated, eq. (3.25) [2].

$$\Delta G_{tr}(Ph_4As^+) = \Delta G_{tr}(Ph_4B^-) = \tfrac{1}{2}\Delta G_{tr}(Ph_4AsPh_4B) \qquad (3.25)$$

This is because, to the extent that eq. (3.25) represents a chemically sensible division of the free energies of transfer of the salt $[Ph_4As^+BPh_4^-]$, we may expect the resulting ionic values to be indicative of the actual changes in solvation energy of the individual ions.*,†

The build-up of a set of individual free energies of transfer of ions based on eq. (3.25) is straightforward. Thus, $\Delta G_{tr}(BPh_4^-)$ follows from the application of eq. (3.25) to the change in solubility of Ph_4AsBPh_4. Then, for example, from the solubility of $KBPh_4$ in solvents we obtain $\Delta G_{tr}(K^+) + \Delta G_{tr}(BPh_4^-)$, and hence $\Delta G_{tr}(K^+)$. In a similar manner ΔG_{tr} values for any number of ions follows: e.g., $\Delta G_{tr}(KCl)$ combined with $\Delta G_{tr}(K^+)$ gives $\Delta G_{tr}(Cl^-)$ from which we may derive $\Delta G_{tr}(H^+)$ using cell 2.17, and $\Delta G_{tr}(Na^+)$ from the solubility of NaCl, etc.

Tables 3.5–3.7 list ΔG_{tr}-values for the transfer of ions from water to alcohols, formamide, and aprotic solvents, respectively. ΔG_{tr}-values for ions in formamide are reported separately; it is formally a protic solvent because of its relatively acidic NH-protons, but the presence of the carbonyl group makes it is significantly more effective than the alcohols at solvating cations.

Considering first the anions, the most obvious conclusion from the results in Tables 3.5–3.7 is *that the transfer of anions from water is almost universally unfavourable.* The effects are particularly striking for high-charge-density

*The logic inherent in eq. (3.25) is that charges on these large and strongly shielded ions are sufficiently buried and dispersed so as to exclude any significant, charge-specific contributions to the total solvation energy of the ions

†For a discussion of the absolute free energy of the proton in different solvents relative to the gas-phase, see Himmel, D.; Sascha, K.G., Leito, I., Krossing, I. *Angew. Chem. Int. Ed.,* 2010, *49*, 6885

Table 3.5 Free energies of transfer of ions[a] from water to alcohols at 25°C[b]

						ΔG_{tr}/kJ mol^{-1}			
Ion	MeOH	EtOH	n-PrOH	n-BuOH	Ion	MeOH	EtOH	n-PrOH	n-BuOH
H^+	10.4	9.1			F^-	17.0	27.0		
Li^+	5.0	11.0	11.3		Cl^-	13.0	20.0	25.5[c]	29.2[c]
Na^+	8.0	14.0	16.8	19.8	Br^-	11.0	18.0	21.9[c]	28.7[c]
K^+	10.0	16.0	17.7[b]	19.8	I^-	7.0	13.0	19.2	22.1
Rb^+	10.0	16.0	19.3	22.6	OAc^-	16.0	36.9		
Cs^+	9.0	15.0	17.4	18.5	BzO^-	7.0			
Ag^+	7.2				N_3^-	9.0	17.0		
Me_4N^+	6.0	11.0	10.6	12.1	CN^-	9.0	20.0		
Et_4N^+	1.0	6.0	4.8	7.3	CNS^-	6.0	13.0		
Pr_4N^+	−5.0	−6.0	−6.4	−6.7	ClO_4^-	6.0	10.0	17.3	21.5
Bu_4N^+	−21.0	8.0	−16.8	−11.7	Pic^-	−6.0	−1.0		
Ph_4As^+	−24.0	−21.0	−25.2	−20.1	BPh_4^-	−24.0	−21.0	−25.2	−20.1

[a] Convention, $\Delta G_{tr}(Ph_4As^+) = \Delta G_{tr}(BPh_4^-)$; [b] Ref [2, 6–8, 10]; [c] ΔG_{tr}/kJ mol^{-1} in *i*-PrOH: $H^+ = 11.0$, $K^+ = 22.5$, $OAc^- = 31.0$, $Cl^- = 22.0$, $Br^- = 19.8$; ΔG_{tr}/kJ mol^{-1} in *t*-BuOH: $H^+ = 18.8$, $OAc^- = 38.9$, $Cl^- = 36.3$, $Br^- = 31.2$, Ref [9, 10]

Table 3.6 Free energies of transfer of ions[a] from water to formamide at 25°C[b]

	ΔG_{tr}/kJ mol^{-1}		
Cation	ΔG_{tr}/kJ mol^{-1}	Anion	ΔG_{tr}/kJ mol^{-1}
Li$^+$	−9.6	F$^-$	24.7
Na$^+$	−8.0	Cl$^-$	13.8
K$^+$	−6.3	Br$^-$	11.3
Rb$^+$	−5.4	I$^-$	7.5
Cs$^+$	−7.5	CN$^-$	13.3
Ag$^+$	−22.6	OAc$^-$	20
Ph$_4$As$^+$	−23.9	BPh$_4^-$	23.9

[a] Convention, $\Delta G_{tr}(\text{Ph}_4\text{As}^+) = \Delta G_{tr}(\text{BPh}_4^-)$; [b] Ref [9–11]

Table 3.7 Free energies of transfer of ions[a] from water to aprotic solvents[b] at 25°C[c]

	ΔG_{tr}/kJ mol^{-1}								
Ion	DMF	DMAC	NMP	DMSO	Me$_2$CO	MeCN	PC	MeNO$_2$	PhNO$_2^d$
H$^+$	−14.8		−18.9	−19.4	14.5	44.8	50		
Li$^+$	−15.1			−10.0	5.0	25.0	25.8	48	
Na$^+$	−10.0	−12.1		−13.7	9.0	15.0	15.2	31.6	
K$^+$	−10.0	−11.7	−11	−12.0	3.9	8.0	6.2	15.4	21.0
Rb$^+$	−10.0	−8.0	−8	−10.8	4.0	7.5	4.9	11.1	19.3
Cs$^+$	−11.0		−10	−12.5	4.0	6.0	2.3	5.7	17.8
Ag$^+$	−17.7	−29.0	−19.8	−33.3		−22.8	16.7	24.7	
Me$_4$N$^+$	−7.4		−5.0	−2.0	10.6	3.0		−4.6	4.0
Et$_4$N$^+$	−8.0			−9.0	−11.0	−7.0		−10.3	−4.8
Et$_3$NH$^+$				−5.4					
Pr$_4$N$^+$	−17.0			−19.0	−20.0	−13.0		−20.0	−16.4
Bu$_4$N$^+$	−29.0			−36.9	−32.9	−30.9			
Ph$_4$As$^+$	−38.4	−38.7	−39.0	−36.9	−32.9	−32.9	−35.4	−32.6	−36.0
OH$^-$				109^e					
F$^-$	85			61	60	70	58		
Cl$^-$	47.9	54.9	55.2	40.0	56.4	41.9	39.4	37.7	43.9
Br$^-$	31.7	44.0	40.6	27.0	41.9	30.9	30.3	29.0	36.0
I$^-$	20.0	21	24.3	10.0	25.0	17.0	16.8	18.8	21.9
OAc$^-$	64.9	70	67.1	61.1	64.6	60.9		56.1	
BzO$^-$	48.0					40.5			
PhO$^-$				61.1^e					
PhS$^-$				35.2^e					
N$_3^-$	36.0		31.7	26.0	43.0	34.3	28.3	25.1	
CN$^-$	40.0			34.9	47.9	36.9			
CNS$^-$	18.0		9.0	10.0	30.0	14.0		8.8	
ClO$_4^-$	4.0			−6.0	10.0	2.0	−3.0	4.7	9.8
Pic$^-$	−7.0			14.0	14.0	4.0	−0.2		−3.4
BPh$_4^-$	−38.4	−38.7	−39.0	−36.9	−32.9	−32.9	−35.4	−32.6	−36.0

[a] Convention, $\Delta G_{tr}(\text{Ph}_4\text{As}^+) = \Delta G_{tr}(\text{BPh}_4^-)$; [b] Abbreviations as in Table 1.1; [c] Ref [2, 6–8, 10, 11]; [d] Danil de Namor, A.F., Hill, T. *J. Chem. Soc., Faraday Trans. 1*, 1983, *79*, 2713; [e] Pliego, J.R., Riveros, J.M. *Phys. Chem. Chem. Phys.*, 2002, *4*, 1622

anions, such as OH$^-$, F$^-$, Cl$^-$, N$_3^-$, CN$^-$, OAc$^-$, and PhO$^-$, on transfer to the various *aprotic* solvents, such as DMF, DMSO, and MeCN. The acetate ion, for example, shows an increase in free energy of some 60kJ mol^{-1} (equivalent to 10.5 units in pK$_a$, log $^{aq}\gamma^S$) in a range of aprotic solvents relative to water. The increases are considerably lower in the *protic* solvents methanol and formamide, but become systematically larger in progressing from methanol to butanol. This is not surprising, as the 'solvating' group, –OH, becomes increasingly 'diluted' as the alkyl chain increases.

Highly polarizable anions with significant charge dispersion, such as CNS$^-$, ClO$_4^-$, picrate, and BPh$_4^-$, show, as might be expected, much smaller increases (and in some cases a decrease) in free energy on transfer from water. For these ions, strong dispersion-force interactions in the non-aqueous solvents largely compensates for any loss of hydrogen-bond stabilization.

The behaviour of cations cannot be classified in such a simple manner, but it is qualitatively in line with expectations based on the structures and charge distributions of the different solvents. Water no longer holds a unique position: interactions of cations with non-aqueous solvents may be stronger or weaker than those with water, depending on the polarity and basicity of the solvent. Thus, 'basic' solvents (Chapter 1, Section 1.2), such as DMSO, DMF, NMP, formamide, and particularly hexamethylphosphoric triamide (HMPT), stabilize simple cations, and especially the proton, relative to water, generally in the order H$^+$ > Li$^+$ > Na$^+$ > K$^+$ > Rb$^+$ ≈ Cs$^+$. The order is reversed for solvents that are less basic than water (PC, MeCN, MeNO$_2$, and acetone); smaller cations, and again especially the proton, may be considerably destabilized. Within the series of alcohols there is a general progressive destabilization of simple cations from methanol to butanol.

'Organic' cations, including the larger R$_4$N$^+$, Ph$_4$As$^+$, and BPh$_4^-$, are less stable in water and methanol than in polar aprotic solvents, but show little variation among polar aprotic solvents. The preference of the 'organic' cations for non-aqueous solvents increases with cation polarisability, as illustrated, for example, by the trend among R$_4$N$^+$ ions as the alkyl chain length increases.

3.3.3 Solvation in mixed solvents

Mixed aqueous–organic solvents and, in particular, alcohol–water mixtures are often convenient for synthetic and purification processes. They are, for example, frequently used for the purification and isolation of pharmaceutical actives, and for the separation of acid- or base-sensitive substrates by HPLC. There are clearly a vast number of such combinations and it is not possible to review the results in any detail. A useful principle, however, is that the properties of the solutes in the solvent mixture will be most strongly influenced by the component with which they interact most strongly in the pure solvents— a phenomenon normally referred to as *selective or preferential solvation.*

One obvious consequence of preferential solvation is that in mixed aqueous-organic solvents the major increases in free energy normally occur only when most of the water is removed from the system. This illustrated in Fig. 3.1, which shows the properties of NaCl and HCl in water–acetonitrile and water–methanol mixtures, respectively [12].

Fig. 3.1.
Free energies of transfer of NaCl and HCl from water to MeCN–water and MeOH–water mixtures, respectively

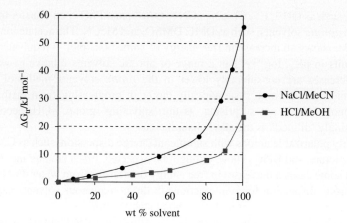

*The detailed shape of such curves (and similar curves for the dependence of acid strength upon solvent composition) depends upon the units used to define the solvent mixtures: wt%, vol%, or mol%. The issue and choice of scale is discussed in Appendix 3.1

In both cases there is a steep rise in free energy beyond 80 wt% of the organic component.*

For the same reason, the presence of small amounts of water in weakly solvating media, such as acetonitrile, can have a strong influence on the properties of electrolytes and the ionization of acids (Chapter 7). In some solvents, such as propylene carbonate, the association of ions with water is sufficiently strong to allow the determination of equilibrium constants for the selective solvation of ions by water in the solvent [13].

A related phenomenon is the solubility in mixed solvents of organic molecules which contain both hydrophobic and hydrophilic components; these may be more soluble in mixed aqueous solvents—for example, THF–water mixtures—than in either of the component solvents separately.

3.4 Solvation of non-electrolytes

The free energies of transfer of various non-electrolytes from water to organic solvents are given in Table 3.8. As noted above, they are derived from measurements of solubility, vapour pressure (Henry's Law coefficients), and partition coefficients between water and other solvents.

Two obvious conclusions may be drawn from the data. First, non-electrolytes are almost universally more stable in non-aqueous media than in water; larger and more polarisable solutes show greater decreases in free energy on transfer from water because of stronger dispersion-force interactions with the non-aqueous solvents [1]. Secondly, changes amongst the various non-aqueous solvents are normally small.

Importantly, though, the solubility of carboxylic acids is enhanced by H-bond formation with suitable solvents, such as DMSO and DMF, compared with solvents, such as MeCN, which are poor H-bond acceptors. Thus an extensive study of the solubility aromatic carboxylic acids and their esters, has shown that while their esters, in common with the majority of non-electrolytes, show little variation amongst a range of non-aqueous solvents, the carboxylic acids are more weakly solvated in solvents such as acetonitrile [3, 14]; compare, for example methyl, 4-bromobenzoate and 4-bromobenzoic acid. On

Table 3.8 Free energies of transfer of non-electrolytes from water[a] to non-aqueous solvents[b] at 25°C

	ΔG_{tr}/kJ mol^{-1} [c,d]					
Substrate	MeOH	DMF	NMP	DMSO	MeCN	Dioxane
ethane	−9.7			−6.3		
ethylene	−7.4	−7.4		−5.2		
CH$_3$Br	−6.8	−8.5				
CH$_3$I	−8.0	−10.8	−12.0	−10.8	−10.3	
t-BuCl	−17.1	−18.2	−16.5			
t-butyl acetate				−12.6		
acetic acid	0.0	1.1	−2.3	−4.6	2.3	
ethyl acetate				−8.4	−9.2	−11.3
benzoic acid	−11.7	−13.0	−12.7		−8.5	
4-bromobenzoic	−15.9	−20.9		−19.4	−11.4	
Me,4-bromobenzoate	−16.7	−21.0		−19.4	−19.1	
4-nitrobenzoic	−13.5	−19.9		−19.3	−9.5	
Me,4-nitrobenzoate	−12.9	−18.5		−17.4	−17.7	
3,4-dichlorobenzoic	−18.8			−24.1	−13.3	
Me,3,4-dichlorobenzoate	−18.3			−20.8	−20.4	
3,4-dimethylbenzoic	−16.7			−20.6	−12.0	
fumaric acid	−5.7			−10.6	5.6	
methyl fumarate	−5.7			−8.4	−2.0	
adipic acid	−5.1	−6.5		−8.1	3.2	
methyl adipate	−8.2			−8.1	−8.2	

[a] Free energy of solution in water, ΔG_s/kJ mol^{-1}: ethylene, 13.1; benzoic acid, 8.87; 4-bromobenzoic acid, 21.12; methyl 4-bromobenzoate, 19.65; 4-nitrobenzoic acid, 18.44; methyl 4-nitrobenzoate, 17.11; 3,4-dichlorobenzoic acid, 22.1; methyl,3,4-dichlorobenzoate, 18.4; fumaric acid, 7.32; methyl fumarate, 4.17; adipic acid, 4.87; methyl adipate, 4.11; [b] Abbreviations as in Table 1.1; [c] Molar concentration scale; [d] Ref. [1, 3, 14]

average, the aromatic esters show a decrease in free energy of 19.1 kJ mol^{-1} on transfer from water to both DMSO and MeCN, whereas the corresponding decreases in free energies of the carboxylic acids are 21.3 and 11.9 kJ mol^{-1}, respectively.

The free energy changes for non-electrolytes, although usually much smaller than those of electrolytes, can contribute up to three orders of magnitude to the observed changes in dissociation constants on transfer from water; for carboxylic acids and phenols, this tends to increase their pK_a-values on transfer from water, whereas the opposite effect occurs for protonated amines and anilines (eqs. (3.4), (3.5)).

3.5 Solvation energies and solvent properties

The solvation energies of cations and anions are expected to be enhanced in solvents which are able to donate or accept electrons, respectively. Indeed, free energies of transfer of cations, such as Na$^+$, correlate well with the solvent Donor Number, DN, which provides a measure of the ability to donate electrons, and those of anions with the solvent Acceptor Number, AN (Chapter 1, Section 1.2.1). This is shown in Fig. 3.2 for the transfer of Na$^+$ and Cl$^-$ among solvents.

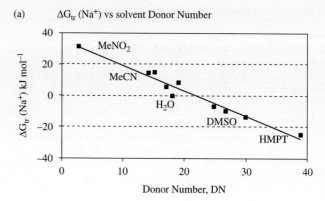

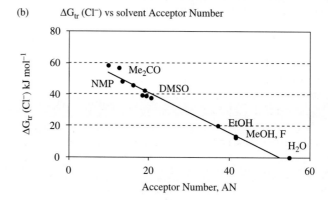

Fig. 3.2.
Influence of solvent Donor Number (a) and Acceptor Number (b) on the free energies of transfer of Na^+ and Cl^-, respectively, from water to non-aqueous solvents (data from Tables 1.1, 3.5–3.7)

The data in Fig. 3.2 provide good confirmation of the importance of electron donor-acceptor interactions in determining the free energies of transfer of simple ions among solvents. Thus, Na^+ becomes progressively more strongly solvated as the Donor Numbers of the solvents increase, whereas Cl^- interacts more strongly with solvents having high Acceptor Numbers. The contrast between protic and aprotic solvents in the solvation of the chloride ion is also apparent in Fig. 3.2(b).

3.6 Solvation and acid strength

In anticipation of a more detailed discussion of dissociation constants in subsequent Chapters, it is instructive at this stage to illustrate the application of the thermodynamic solvation data to the analysis of pK_a changes in different solvents, using the pK_a of acetic acid as an example.

Table 3.9 provides an analysis of the change in pK_a of acetic acid in acetonitrile, dimethylformamide, and methanol relative to water in terms of the free energy changes of the components of the equilibrium. The free energy changes of HOAc, H^+ and OAc^- are expressed in terms of their solvent transfer activity coefficients, $\log{}^{aq}\gamma^S$, eq. (3.7), in order, illustrate more clearly their contribution to the change in pK_a. The overall change in pK_a to be expected then follows from eq. (3.10), i.e.,

Table 3.9 Influence of solvent on the dissociation constant of acetic acid

$$HOAc \rightleftharpoons H^+ + OAc^-$$

	MeCN	DMF	MeOH
$\log{}^{aq}\gamma^S(HOAc)^a$	0.4	0.2	0.0
$\log{}^{aq}\gamma^S(H^+)^a$	7.9	−2.6	1.8
$\log{}^{aq}\gamma^S(OAc^-)^a$	10.7	11.4	2.8
ΔpK_a^b	18.2	8.6	4.6
$pK_a(H_2O)$	4.76	4.76	4.76
$pK_a(S_{calc})^c$	23.0	13.4	9.4
$pK_a(S_{obs})^d$	23.1	13.5	9.7

^{a} $\log{}^{aq}\gamma^S(Y) = \Delta G_{tr}(Y)/2.303RT$, $\Delta G_{tr}(Y)$ from Tables 3.5, 3.7–8; ^{b} Eq. (3.10); ^{c} $pK_a(H_2O) + \Delta pK$; ^{d} Measured values, Chapters 5–7

$$\Delta pK_a = \log{}^{aq}\gamma^S(H^+) + \log{}^{aq}\gamma^S(OAc^-) - \log{}^{aq}\gamma^S(HOAc)$$

Values calculated in this way from the solvation data can be seen to be in good agreement with directly measured pK_a-values.*

The greatest increase in pK_a (18.2 units) when compared with aqueous values occurs on transfer to MeCN, where both H^+ and OAc^- show large increases in free energy. In DMF, the acetate ion is similarly unstable compared with water, but the proton is more strongly solvated. The result is a much smaller increase in pK_a (8.6 units) on transfer from water. Methanol shows a modest increase in pK_a values; both H^+ and OAc^- are solvated more poorly than in water, but the increase in free energy of OAc^- in particular is much lower than in either of the aprotic solvents.

*Note that because the dissociation equilibrium involves the neutral combination of H^+ and OAc^-, the results are, therefore, independent of the convention used to derive the single-ion values of $^{aq}\gamma^S(H^+)$ and $^{aq}\gamma^S(OAc^-)$

3.7 Summary

The discussion may be summarized as follows:

- The absolute solvation energies of ions are very large (several hundred to several thousand kJ mol^{-1}) and dominated by electrostatic effects, but changes among solvents are governed mostly by specific interactions within the first coordination sphere.
- Small, high-charge-density anions, such as hydroxide, carboxylate and halide ions, are strongly stabilized by hydrogen-bond formation in protic solvents. They show large increases in free energy on transfer to aprotic solvents, which may reach 100kJ mol^{-1}, *equivalent to 18 pK-units.*

$$A^- \cdots \overset{\delta+}{H}{-}\overset{}{\underset{\delta-}{O}}\overset{\delta+}{\diagup} \quad > \quad A^- \cdots \overset{\delta+}{H}{-}\overset{R}{\underset{\delta-}{O}}\overset{}{\diagup} \quad \gg \quad A^- \cdots {-}O{-}\overset{CH_3}{\underset{CH_3}{S^+}}$$

(similarly DMF, NMP, etc.)

- Cations, such as the proton and alkali metal ions, are stabilized relative to water in 'basic' solvents, such as DMF, DMSO, and NMP, but have considerably higher free energies in less basic solvents, such as acetonitrile.

$$M^+ \cdots {}^-O-S^+\begin{smallmatrix}CH_3\\ \\CH_3\end{smallmatrix} \quad > \quad M^+ \cdots {}^{\delta-}O\begin{smallmatrix}H^{\delta+}\\ \\H^{\delta+}\end{smallmatrix} \quad > \quad M^+ \cdots \overset{\delta-}{N}\equiv\overset{\delta+}{C}-CH_3$$

(similarly DMF, NMP, etc.)   (similarly PC, sulfolane, etc.)

- In mixed-aqueous solvents, the phenomenon of preferential solvation means that the major changes in free energy occur only as the better solvent (usually water) is significantly depleted.
- Semi-empirical measures of electron donor/electron acceptor properties, such as Donor Numbers, Hydrogen-bond Basicity, and Acceptor Numbers, provide a means of correlating and qualitatively predicting changes in free energies amongst solvents.
- Non-electrolytes are almost universally more stable in non-aqueous media, but the effects are generally much smaller than those for ions. Larger, 'organic' electrolytes, such as those involving alkylamonium ions, behave more like non-electrolytes in terms of their solvation behaviour.
- Analysis of the changes in acid dissociation constants and acid–base equilibria among solvents in terms of the changes in free energy of the individual species involved can provide an enhanced understanding of solvent effects on acid–base equilibria.

Appendix 3.1 Composition of mixed solvents

Free energies of transfer of electrolytes and dissociation constants of acids (Chapter 5, 8) in mixed solvents, most commonly in practice mixed-aqueous solvents, frequently show strong evidence of preferential solvation as, for example in the data for NaCl in acetonitrile–water and methanol–water mixtures, Section 3.3.3, Fig. 3.1.

There are three ways of expressing the composition of liquid mixtures, the choice of which can influence the interpretation of preferential solvation: mole fractions, weight fractions and volume fractions. The latter two are very similar, the differences depending only upon the relative densities of the component solvents, but differ strongly from the mole fraction scale *when the two solvent components have very different molecular weights*; this is typically the case when water is one of the components. Table A3.1 and Fig. A3.1 show the free energy of transfer of NaCl between water and water-MeCN mixtures according to the different composition scales.

All three curves exhibit evidence of preferential solvation by water in the mixtures, in that the increase in free energy is more marked at low water levels, but the effect appears to be more pronounced when composition scales are expressed in terms of vol% or wt%, rather than mol%.

An exactly analogous situation obtains for the dependence of acid dissociation constants upon solvent composition, discussed in subsequent chapters (Chapters 5 and Chapter 8).

Table A3.1 Free energy of transfer of NaCl from water to water-acetonitrile mixtures at 25°C[a]

Vol% MeCN	Wt% MeCN	Mol% MeCN	ΔG_{tr}/kJ mol^{-1}
0	0	0	0
10	7.9	3.6	1.13
20	16.4	7.9	2.13
40	34.3	18.6	5.19
60	54.1	34.1	9.33
80	75.8	57.9	16.5
90	87.6	75.5	29.4
95	93.8	86.9	40.6
100	100	100	56.1

[a] Ref [12]

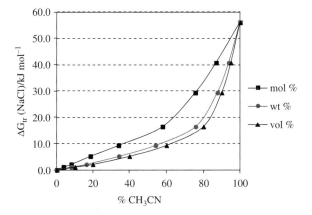

Fig. A3.1.
Free energy of transfer of NaCl from water to acetonitrile-water mixtures at 25°C

The most satisfactory choice for the composition scale as a basis for interpreting solvation changes is that of the *volume fraction scale* (or, almost equivalently, the weight fraction scale), which is the analogue of the molar concentration scale. It reflects more accurately the changes in *chemical/physical interactions* between the solvent components and the solutes that accompany changes in solvent composition. The mole fraction scale, by contrast, additionally includes a strong contribution arising purely from differences in the size of the solvent molecules, irrespective of the nature of the interactions. Thus, for example, in 60 vol% acetonitrile, the *volume concentration* of water molecules is reduced by 60%, compared with pure water, whereas there is a decrease of only 38% in the mole fraction of water molecules. It is the former which governs the free energy of interaction between the ions and water molecules and hence the free energies and dissociation constants of the acid–base species.*

In practice we have chosen throughout this text to express solvent composition in terms of wt%. As an indicator of preferential solvation, it is essentially equivalent to the vol% scale (Fig. A3.1), but the preparation of mixtures by weight is more convenient than by volume, especially on a large scale; use of the wt% scale also avoids the minor issue of non-zero volumes of mixing of the solvent components.

*A similar argument applies to the properties of solvent mixtures in general. Thus, mixtures are normally defined as ideal when they obey Raoult's law, i.e., the vapour pressure of each component is proportional to its *mol fraction* in the mixture. If we are interested primarily in the influence of the interaction energies between the components on their properties, however, the mol fraction scale is less than satisfactory. Thus, even in cases where the interaction energies between the components are identical, mixtures of molecules of different size exhibit non-ideal behaviour when expressed as mole fractions. An extreme example of this occurs in the thermodynamics of polymer-solvent mixtures, where volume fraction statistics are used in order to avoid this problem.

References

[1] Parker, A. J. *Chem. Rev.,* 1969, *69,* 1
[2] Kolthoff, I. M., Chantooni, M. K. *J. Phys. Chem.,* 1972, *76,* 2024
[3] Chantooni, M. K., Kolthoff, I. M. *J. Phys. Chem.,* 1974, *78,* 839
[4] Burgess, J. 'Metal Ions in Solution', 1978, Ellis Horwood, Ch 7
[5] Kebarle, P. *Annu. Rev. Phys. Chem.,* 1977, *28,* 445
[6] Johnsson, M, Persson, I. *Inorg. Chim. Acta,* 1987, *127,* 15
[7] Abraham, M. H., Zhao, Y. H. *J. Org. Chem.,* 2004, *69,* 4677
[8] Marcus, Y. *Pure & Appl. Chem.,* 1983, *55,* 977
[9] Kolthoff, I. M., Chantooni, M. K. *J. Phys. Chem.,* 1979, *83,* 468
[10] Abraham, M. H., Acree, W. E. *J. Org. Chem.,* 2010, *75,* 1006
[11] Cox, B. G., Hedwig, G. R., Parker, A. J., Watts, D. W. *Aust. J. Chem.,* 1974, *27,* 477
[12] Cox, B. G., Waghorne, W. E. *Chem. Soc. Rev.,* 1980, *9,* 381
[13] Butler, J. N., Cogley, D. B., Grunwald, E. *J. Phys. Chem.,* 1971, *75,* 1477
[14] Chantooni, M. K., Kolthoff, I. M. *J. Phys. Chem.,* 1973, *77,* 527

Determination of Dissociation Constants

4

Methods of determining dissociation constants in aqueous solution are well established, and analogous methods are also used in non-aqueous media. They are mostly based on the measurement of pH in solutions of known acid and base concentrations, or measurement of the ratios of acid and base in solutions of fixed pH.

Fundamental to the dissociation constant in any solvent, however, is the pH-scale upon which the acidity is based, and we begin by considering pH-scales in aqueous and non-aqueous solvents.

4.1 pH-scales

4.1.1 pH in aqueous media

pH is defined by eq. (4.1), in which the hydrogen ion activity, a_H, is related to its concentration by eq. (4.2) [1, 2].

$$pH = -\log a_H \quad (4.1)$$

$$a_H = \gamma_H [H^+] \quad (4.2)$$

In eq. (4.2), γ_H is the *molar* activity coefficient of the hydrogen ion in the solution in question (Section 4.2). It measures primarily the influence of interactions with other solutes, predominantly electrostatic interactions with the ions, on the free energy of the hydrogen ion.* The reference state for the activity coefficient is infinite dilution in water, where the only interactions experienced by the proton are those with the solvent. Thus, as the solution concentration approaches zero, $a_H \rightarrow [H^+]$ and hence $\gamma_H \rightarrow 1$, so that, in dilute solution:†

$$pH \approx -\log[H^+] \quad (4.3)$$

Traditionally, the pH of a solution, X, was measured using a standard hydrogen electrode in a cell such as (4.4), in which the reference half-cell comprises an Ag/AgCl electrode in 0.1M KCl; the connection between the two half-cells is provided by a salt bridge, typically comprising saturated KCl.

$$Pt, H_2(g)| \text{soln.} X| \text{salt bridge}| 0.1M KCl| AgCl, Ag \quad (4.4)$$

*Other interactions with solute ions or molecules, such as ion-dipole interactions, tend to be shorter-range but can become important at higher solution concentrations (Section 4.2)

†The standard state is a hypothetical (ideal) solution of concentration 1M, in which the interactions experienced by the proton are the same as at infinite dilution [1]. It is independent of temperature. This is analogous to the situation for gases, for which the standard state is a hypothetical (ideal) gas at a pressure of 1atm.

The half-cell reactions are given by eq. (4.5).

$$AgCl + e \rightarrow Ag + Cl^- \text{ (in 0.1M KCl)} \quad (4.5a)$$

$$H^+ \text{(in soln. X)} + e \rightarrow \tfrac{1}{2} H_2(g) \quad (4.5b)$$

The overall cell reaction, eq. (4.6), includes, in addition to the two half-cell reactions, a term arising from the transfer of ions within the cell.

$$\tfrac{1}{2} H_2(g) + AgCl \rightarrow Ag + H^+(\text{soln. X})$$
$$+ Cl^-(0.1M\ KCl) \pm \text{ion transfer} \quad (4.6)$$

Ion transfer is an integral part of the cell reaction; it is required to maintain electrical neutrality in the two half-cells. Thus, the passage of one electron generates Cl^- in the 0.1M KCl solution at the cathode ($AgCl + e \longrightarrow Ag + Cl^-$) and H^+ in solution X at the anode ($\tfrac{1}{2} H_2 \longrightarrow H^+ + e$) which, in the absence of any balancing factors, would potentially result in a build-up of net negative and positive charges, respectively, in the anodic and cathodic compartments. This is avoided by a matching transfer of ions within the cell, which is achieved predominantly by transfer of K^+/Cl^- from the concentrated salt bridge into the reference half-cell and solution X, respectively.*

*High concentrations of ions in the salt bridge relative to those in the two half-cell solutions are used to ensure that under most conditions that the bulk of the current is carried by transport of the salt bridge ions, so that E_j is independent of the nature of solution X

The potential of cell 4.4, E_X, is given by eq. (4.7), in which E^o is the standard cell potential, E_j is the junction potential associated with the ion transfer, F is the Faraday constant, and a_H is the activity of the hydrogen ions in solution X.

$$E_X = E^o - (RT/F) \ln a_H + E_j$$
$$= E^o + (2.303RT/F)\text{pH}(X) + E_j \quad (4.7)$$

Hence,

$$\text{pH}(X) = (E_x - E^o - E_j)(F/2.303RT) \quad (4.8)$$

In water, there is a wide range of buffer solutions of known pH and these may be used to eliminate the unknown E^o and E_j terms in eq. (4.8). Thus, for a standard or reference (buffer) solution of known pH, pH(R), its pH and cell potential, E_R, are related by an analogous equation, (4.9), provided that the junction potential, E_j, is the same as that for solution.†

†This assumption, which relies on ion-transport being restricted to the salt bridge ions, generally holds in the pH range 2–12; outside this range, some of the ion transfer may involve either H^+ (low pH) or OH^- (high pH) and corrections may be required

$$\text{pH}(R) = (E_R - E^o - E_j)(F/2.303RT) \quad (4.9)$$

Subtraction of eq. (4.9) from eq. (4.8) eliminates E_j and E^o and gives eq. (4.10), *the practical basis for the measurement of pH.*

$$\text{pH}(X) = \text{pH}(R) + (E_x - E_R)(F/2.303RT) \quad (4.10)$$

In modern practice, the standard hydrogen electrode is normally replaced by a glass electrode, which is also directly responsive to hydrogen ion activity and is much more convenient to use, cell (4.11).‡

‡The reference electrode is typically included within the glass electrode

$$\text{Glass electrode}\,|\,\text{soln.X}\,|\,\text{salt bridge}\,|\,0.1M KCl\,|\,AgCl, Ag \quad (4.11)$$

Alternative reference electrodes to Ag/AgCl in 0.1 K KCl may also be used, but *in all cases, eq. (4.10) remains the basis for pH measurements in aqueous solution.*

An inescapable consequence of the conventional scale for pH and hydrogen-ion activity is that *a separate pH scale is required for each temperature*. The standard state, which is independent of temperature, corresponds to setting the standard potential of the hydrogen electrode to zero at each temperature. Thus the pH at one temperature has no quantitative meaning relative to that at another temperature, and therefore the measurement of hydrogen electrode (or, equivalently, glass electrode) potentials at two different temperatures can give no exact comparison between the 'absolute' hydrogen-ion activities at these temperatures.

4.1.2 pH in non-aqueous media

pH-scales may be defined in mixed-aqueous or non-aqueous media in an exactly analogous manner to that for aqueous solution; i.e, via eq. (4.12), in which pH^S and $a_H{}^S$ are the pH and hydrogen-ion activities, respectively, in solvent S, *referred to infinite dilution in solvent S*.

$$pH^S = -\log a_H{}^S \qquad (4.12)$$

Such pH-scales are entirely analogous to those in water and are equally valid as a measure of acidity and dissociation constants in non-aqueous solvents.

It is important to note that a separate pH-scale is required for each solvent, as any given pH-scale is valid only for the solvent (and temperature) to which it refers. Furthermore, *the pH in any solvent S bears no exact quantitative relationship to the pH in water*. Thus, for example, pH 4 in methanol refers to the pH at which a dilute solution of an acid of $pK_a = 4$ in methanol will be 50% ionized, etc., but it is not possible to say whether in absolute terms it is more or less acidic than an aqueous solution of pH 4.

pH^S of a solution X can be measured using cells equivalent to (4.4) and (4.11), in an exactly manner similar to those discussed above, but with solvent S replacing water in the unknown and standard solutions, eq. (4.13).

$$pH^S(X) = pH^S(R) + (E_x - E_R)(F/2.303RT) \qquad (4.13)$$

Furthermore, the relationship between pH^S, pK_a^S and the concentrations of acid, A, and base, B, has the same form, eq. (4.14), as its aqueous counterpart, eq. (2.10).

$$pH^S = pK_a{}^S + \log_{10} \frac{[B]}{[A]} \qquad (4.14)$$

4.2 Influence of solution concentration: activity coefficients

Dissociation constants, when defined in terms of concentrations, normally vary with the total solution concentration, and allowance must be made for this in accurate work. The dependence upon concentration arises primarily because of interactions between the solution components, particularly a net electrostatic attraction between ions, which become stronger as the solutions increase in concentration.§ Thus, for a neutral acid HA, K_a is more strictly defined by

§ In sufficiently concentrated solutions the nature (activity) of the solvent is also altered by the presence of the solutes

eq. (4.15), in which γ_i represents the activity coefficient of species i, and [H$^+$] represents the molar concentration of H$^+$, etc.

$$\text{HA} \rightleftharpoons \text{H}^+ + \text{A}^-: \quad K_{HA} = \frac{[\text{H}^+][\text{A}^-]}{[\text{HA}]} \cdot \frac{\gamma_H \gamma_A}{\gamma_{HA}} \quad (4.15)$$

For simple, monovalent electrolytes, we may define a mean ionic activity coefficient, $\gamma_\pm$, as in eq. (4.16), where γ_+ and γ_- are the activity coefficients of the cation and anion, respectively.

$$\gamma_\pm = (\gamma_+ \gamma_-)^{1/2} \quad (4.16)$$

Furthermore, for all but the most concentrated solutions, the activity coefficients for neutral species remain close to unity, i.e., $\gamma_{HA} \approx 1$, and hence, to a good approximation, we may write eq. (4.17) for K_{HA},

$$K_{HA} = \frac{[\text{H}^+][\text{A}^-]}{[\text{HA}]} \cdot (\gamma_\pm)^2 \quad (4.17)$$

For cationic acids, BH$^+$, the situation is somewhat simpler, as K_{BH} values based on solution concentrations are largely independent of concentration, eq. (4.18).

$$\text{BH}^+ \rightleftharpoons \text{H}^+ + \text{B}: \quad K_{BH} = \frac{[\text{H}^+][\text{B}]}{[\text{BH}^+]} \quad (4.18)$$

This is because the ionic activity coefficients cancel, and $\gamma_B \approx 1$ under most circumstances.

The mean activity coefficient of an electrolyte, cation charge z_+, anion charge z_-, is best calculated using one of a number of semi-empirical equations based on the theory of Debye and Hückel, the most successful of which, eq. (4.19), is due to Davies [3].*

*The expression for activity coefficient for ion i, γ_i, is identical to that in eq. (4.19), except that z_+z_- is replaced by the square of the ionic charge, z_i^2 [3]

$$\log \gamma_\pm = -Az_+z_- \left(\frac{\sqrt{I}}{1 + \sqrt{I}} - 0.3I \right) \quad (4.19)$$

In eq. (4.19), I is the ionic strength, defined by eq. (4.20), where the summation extends over all anions and cations in solution, and c_i is the concentration of ion i.

$$I = \tfrac{1}{2}\Sigma c_i z_i^2 \quad (4.20)$$

†The term $-Az_+z_-\sqrt{I}$ accounts for the effect of long-range coulomb forces only. At most practical concentrations, allowance must also be made for (a) the finite size of the ions, resulting in the inclusion of the denominator $(1 + \sqrt{I})$ in the first term in eq. (4.19), and (b) a variety of short-range ion–ion and ion–solvent interactions, all of which give rise to an approximately linear variation of log $\gamma_\pm$ with concentration (ionic strength), represented by the final term in eq. (4.19)

For the common case of acid–base equilibria referring to 1:1 electrolytes, I is simply equal to the electrolyte concentration: $I = c = c_+ = c_-$.

The constant, A, is given by eq. (4.21), where ε_r is the dielectric constant of the solvent.

$$A = 1.8246 \times 10^6/(\varepsilon_r T)^{3/2} \quad (4.21)$$

At low ionic strength (≤ 0.01M in high dielectric media), eq. (4.19) reduces to the familiar Debye–Hückel limiting law, eq. (4.22).†

$$\log \gamma_\pm = -Az_+z_-\sqrt{I} \quad (4.22)$$

Table 4.1 Ionic activity coefficients in solvents at 25°C[a]

Solvent[b]	ε_r	A[c]	$\gamma_\pm$ I = 0.001 M	$\gamma_\pm$ I = 0.01 M	$\gamma_\pm$ I = 0.1 M
HCONH$_2$	109.5	0.309	0.979	0.940	0.861
H$_2$O	78.5	0.512	0.966	0.902	0.782
PC	64.9	0.678	0.953	0.871	0.721
DMSO	48.9	1.037	0.931	0.811	0.605
MeCN	37.5	1.546	0.897	0.731	0.474
DMF	36.7	1.569	0.895	0.724	0.462
MeOH	32.6	1.906	0.875	0.679	0.397
NMP	31.5	2.007	0.869	0.667	0.378
Acetone	20.6	3.794	0.767	d	d
i-PrOH	19.9	3.996	0.757	d	d
THF	7.6	16.93	0.306	d	d

[a] Calculated using eq. (4.19); [b] Abbreviations as in Table 1.1; [c] Eq. (4.21); [d] High level of ion association (see text)

Eq. (4.19) has been tested extensively in aqueous solution and shown to reproduce measured activity coefficients to within a few percent up to an ionic strength of 0.5M. It has also been widely and successfully used in polar non-aqueous media, provided due allowance is made for ion association where appropriate.

Table 4.1 lists activity coefficients, $\gamma_\pm$, in solvents of varying dielectric constant at ionic strengths of 0.001M, 0.01M and 0.1M. Activity coefficients in other solvents may be calculated using eq. (4.19) combined with ε_r from Table 1.1 or standard compilations.

An example of the influence of ionic strength on the observed dissociation constant of an acid is given in Appendix 4.1 for the dissociation of acetic acid in the presence of increasing concentrations of sodium chloride.

For the higher-dielectric, polar solvents, $\gamma_\pm$ values are close to 1 ($\log \gamma_\pm \sim 0$), except at high concentrations, and deviations from ideality make only a modest contribution to acid–base equilibria. Electrolytes in low-dielectric media, such as acetone, i-PrOH, and especially THF, exhibit highly non-ideal behaviour, even at low concentrations.

In practice, dissociation constants in non-aqueous media are mostly measured at low concentrations ($\leq 10^{-3}$M) in order to minimize the effects of general *ion-association* (ion-pair, ion-triplet formation) and of more specific forms of *hydrogen-bonded, ion–molecule association*; thus activity coefficients are often close to unity.

4.3 Ion association

Ion-pair formation between cation, M$^+$, and anion, X$^-$, eq. (4.23), is especially prevalent in solvents of low dielectric constant [4–7].

$$M^+ + X^- \xrightleftharpoons{K_{IP}} (M^+X^-) \qquad (4.23)$$

In THF, for example, K_{IP} values are typically in the range 10^5–10^6 M^{-1}. This means that even at electrolyte concentrations as low as 10^{-5}M, ion-pair

Table 4.2 Ion-pair formation for salt MX

	% ion-pair formation				
[MX]/M	$K_{IP} = 10\,M^{-1}$	$K_{IP} = 10^2\,M^{-1}$	$K_{IP} = 10^3\,M^{-1}$	$K_{IP} = 10^4\,M^{-1}$	$K_{IP} = 10^5\,M^{-1}$
10^{-1}	38.2	73.0	90.5	96.9	99.0
10^{-2}	8.4	38.2	73.0	90.5	96.9
10^{-3}	1.0	8.4	38.2	73.0	90.5
10^{-4}	0.1	1.0	8.4	38.2	72.9
10^{-5}	0.01	0.1	1.0	8.4	38.2

*$K_{IP} = x/(a-x)^2$, where a = stoichiometric concentration of MX, x = ion-pair concentration; x can be calculated by rearrangement to a standard quadratic equation

formation is important, and higher forms of association, such as ion-triplets, become significant at concentrations of around 10^{-3} M. Contrasting this are solvents at the top of Table 4.1, DMSO to $HCONH_2$, in which ion-pair formation constants are typically less than $100\,M^{-1}$ and are often too low to measure. Solvents of intermediate polarity, CH_3CN to NMP, Table 4.1, exhibit ion-pair formation constants of between $100\,M^{-1}$ and $1000\,M^{-1}$, which means that ion-pair formation becomes significant at concentrations of around 10^{-3} M and above.

The dependence of the extent of ion-pair formation at different concentrations of salt, MX, upon K_{IP} can be readily calculated from a simple quadratic equation,* and representative results are shown in Table 4.2.

Determinations of dissociation constants using spectrophotometric techniques (see Section 4.5.2) are typically performed at concentrations $\leq 10^{-4}$ M, so are normally only influenced by ion-pair formation in solvents of very low dielectric constants, but potentiometric measurements often involve concentrations of 10^{-3} M or above. Practical applications of acids and bases in synthesis or salt formation are performed at much higher concentration (> 0.1 M) and are therefore usually accompanied by extensive ion-pair formation in non-aqueous media—even those with relatively high polarities.

Appendix 4.2 shows the relationship between pK_a-determination and ion-pair formation.

†Also known as *homoconjugation*

4.4 Homohydrogen-bond formation†

Carboxylic acids and phenols, especially in solvents in which anions are poorly solvated, such as the polar aprotic solvents (Section 1.2), tend to form strong hydrogen-bond pairs, as in eq. (4.24) for carboxylic acids, and similarly for phenols [6–10].

$$RCO_2H \xrightleftharpoons{K_a} H^+ + RCO_2^-$$

$$RCO_2H + RCO_2^- \xrightleftharpoons{K_{AHA}} RCO_2^- \cdots HO_2CR \qquad (4.24)$$

Representative values of the association constant are given in Table 4.3.

K_{AHA} typically increase in the sequence DMSO < DMF ~ NMP < CH_3CN ~ PC, the order reflecting primarily the ability of the solvent to stabilize the acid by H-bond formation (highest for DMSO) and hence to

Table 4.3 Homohydrogen-bond constants, K_{AHA}, in aprotic solvents [a,b]

	K_{AHA}/M^{-1}				
Acid	DMSO	DMF	NMP	MeCN	PC
methane sulfonic				3×10^3	2.5×10^3
benzoic	6×10^1	2.5×10^2	1×10^3	4×10^3	9×10^3
acetic	3×10^1	4×10^2		4.7×10^3	
phenol	4×10^2		1×10^4	1.1×10^4	
4-NO_2-phenol	4×10^1	2×10^2	6.3×10^2	5×10^3	

[a] Abbreviations as in Table 1.1; [b] Ref [6–10]

reduce the tendency to associate with the anion. They are considerably smaller for *o*-substituted phenols, presumably for steric reasons. In particular, they are usually negligible for picric acid, which is often used to provide standard buffer solutions for calibration of glass electrodes in non-aqueous media (Section 4.5.1).

Similar forms of hydrogen-bonded association may also be observed between amines and ammonium ions, but the magnitudes of the association constants, K_{BHB}, are significantly smaller [9–13]. For example, in acetonitrile, K_{BHB} for primary amines are typically around $20 \, M^{-1}$ and are less than $10 \, M^{-1}$ (and often undetectable) for secondary and tertiary amines [13].

Homohydrogen-bond formation has no effect on the pH when the acid is exactly half-neutralized; it removes both species equally and therefore the ratio of free anion to free cation remains unaltered at $[A^-]/[HA] = 1$. At lower or higher ratios of A^-/HA, however, the pH will be lower or higher, respectively, than that predicted in the absence of association. For example, at high ratios of $[A^-]/[HA]$, the reduction in $[HA]$ through formation of AHA^- will be proportionately much greater than that of $[A^-]$, and hence the ration $[A^-]/[HA]$ will increase, giving rise to a corresponding increase in measured pH, eq. (4.25).

$$pH^S = pK_a^S + \log_{10} \frac{[A^-]}{[HA]} \qquad (4.25)$$

The influence of homohydrogen bond formation on a typical pH-ionization curve for an acid, HA, may be determined by solving eq. (4.24) for individual values of $[HA]$ and $[A^-]$ at different total solution concentrations. Fig. 4.1 shows the output of such a calculation, using as an example an acid with $pK_a = 7.0$ and $K_{AHA} = 10^3 \, M^{-1}$. Also included is the 'normal' curve, i.e., that for the case in which homohydrogen bonding is absent, $K_{AHA} = 0$.

The curves pass through a common point at half neutralization ($pH = pK_a = 7$), but deviate increasingly at high or low degrees of ionization. The deviations become more pronounced as the total concentration of the acid increases, because of the second-order nature of the association. For an association constant, $K_{AHA} = 10^3 \, M^{-1}$, 26% of the acid is associated at half-neutralization at a concentration of $10^{-3} M$, and this rises to 86% at 0.1 M total acid.

Methods for determining K_{AHA} values are outlined in Appendix 4.3.

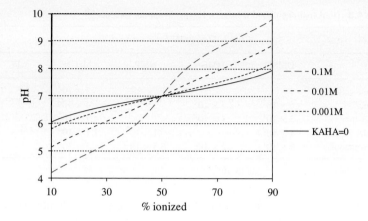

Fig. 4.1.
Dependence of the observed pH versus ionization curves for acid HA, $pK_a = 7$, $K_{AHA} = 10^3 M^{-1}$ (eq. (4.24)), upon total acid concentrations, $[HA] = 10^{-3}-10^{-1} M$: the pH curve in the absence of association ($K_{HAA} = 0$) is included for comparison

4.5 Experimental methods for the determination of dissociation constants

Dissociation constants in non-aqueous media are measured by methods analogous to those used in water. They mostly follow from a simple consideration of the relationship between pH, pK_a and ratio of base to acid (from rearrangement of eq. (4.14)):

$$pK_a^S = pH^S - \log_{10}\frac{[B]}{[A]}$$

For weak acids ($pK_a \geq\sim 6$), the experimental methods used fall essentially into two categories. The first is to fix the ratio [B]/[A], typically via an acid–base titration in which the ratio is known at each point in the titration, and to then determine the pH either by means of an electrochemical cell, eq. (4.11), or by using an acid-base indicator of known pK_a whose composition (pH) can be determined spectrophotometrically. The second is to fix the pH by using a buffer of known pK_a, and then to determine the ratio [B]/[A] at low concentration in the buffer, using (most commonly) ultraviolet or visible spectrophotometry.

In all cases, however, the measurements must be anchored to a solution of known pH (in order to calibrate the glass electrode) or to an indicator acid of known pK_a. Solutions of known pH may be generated using either strong acids, which are fully dissociated under the experimental conditions, or a reference acid whose pK_a in the solvent is known or can be conveniently measured.

For sufficiently strong acids ($pK_a \leq\sim 6$), comparable concentrations of each of the species (e.g., HA, H^+ and A^-) exist at equilibrium in dilute solutions. It is then possible to determine K_a by direct determination of the concentrations of the individual species using spectrophotometric (Section 4.5.2) or conductimetric measurements.

For poorly soluble acids, the increase in solubility of the acid with increasing pH can be used to determine its pK_a and likewise for insoluble bases, the decrease in solubility with increasing pH.*

---

*For an acid, HA, solubility, S_o, the total solubility, $[HA]_T = S_o + [A^-]$, as a function of $[H^+]$ is given by $[HA]_T = S_o(1 + K_a/[H^+])$. Hence, measurement of $[HA]_T$ at known $[H^+]$ (pH) will give K_a, provided S_o is known. Similarly, for base B, solubility, S_o, $[B]_T = S_o + [BH^+] = S_o(1 + [H^+]/K_a)$: Atherton, J. H.; Carpenter, K. J. 'Process Development: Physicochemical Concepts', OUP, 1999

Experimental methods for the determination of dissociation constants

4.5.1 Potentiometric titration using a glass electrode

Potentiometric titration with a glass electrode has been widely used for the determination of dissociation constants in non-aqueous media and especially in mixed aqueous-organic solvents. The basis for the determination of the pH of a solution X is cell (4.11), in which the pH and cell potential, E_X, may be expressed by eq. (4.26) (derived from eq. (4.10), in which the calibration potential, E^{cal}, is given by eq. (4.27) and E_R is the cell potential for a solution of known pH, pH(R), *in the solvent in question*.

$$\text{pH}(X) = (E_X - E^{cal})(F/2.303RT) \quad (4.26)$$

$$E^{cal} = E_R - (2.303RT/F)\text{pH}(R) \quad (4.27)$$

Furthermore, the constant $2.303RT/F$ has the value 59.1 mV at 25°C and so eq. (4.26) reduces to eq. (4.28), in which E is measured in mV.

$$\text{pH}(X) = (E_X - E^{cal})/59.1 \quad (4.28)$$

In aprotic solvents, solutions of known pH are most conveniently produced using a buffer solution of an acid that does not have a strong tendency towards homohydrogen-bond formation. The reference acid normally has a relatively low pK_a and suitable spectral properties such that its pK_a can be readily determined from absorbance measurements in dilute solution (Section 4.5.2).

Commonly used reference acids fulfilling these criteria are shown in Scheme 4.1.

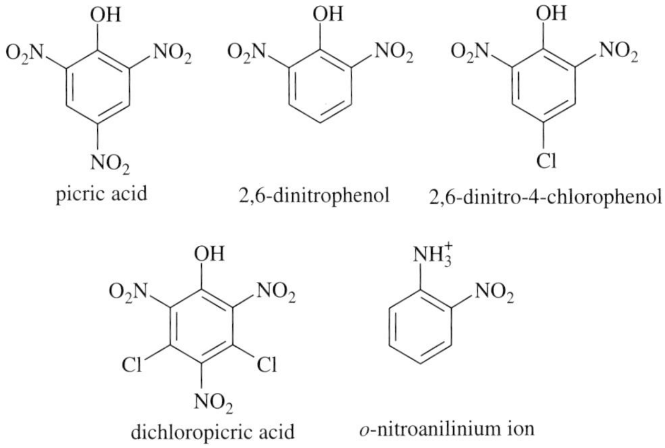

Scheme 4.1.
Acids used in the calibration of glass electrodes in aprotic solvents

Fig. 4.2 shows an example of an extensive calibration plot established in acetonitrile to confirm the validity of eq. (4.26) [14]. It was derived using buffers prepared from the acids 3,5-dichlorobenzenesulfonic acid (3, 5-Cl$_2$-BSA), $pK_a = 6.23$ and methane sulphonic acid (MSA), $pK_a = 10.0$, and their respective Et$_4$N$^+$ salts, and o-nitroanilinium (o-NO$_2$-An, $pK_a = 4.85$) perchlorate and o-nitroaniline mixtures.

In most cases, a simple calibration using only one of the acids in Scheme 4.1 suffices.

Kolthoff and co-workers give full details of the procedure in solvents such as DMSO, DMF, MeCN, and PC [8, 9, 11, 14, 15].

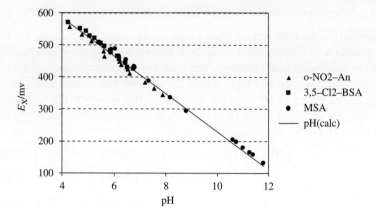

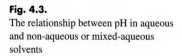

Fig. 4.2.
Calibration of a glass electrode in acetonitrile: the full line corresponds to pH(calc) = $(821 - E_X)/59.1$

Having calibrated the glass electrode, pH-values measured during an acid–base titration can be used in the normal way to derive the desired pK_a via eq. (4.25), or more precisely, eq. (4.29), or its equivalent for a cationic acid. Log $\gamma_\pm$ can be calculated from the Davies eq. (4.19), as discussed above.

$$pH^S = pK_a^S + \log([A^-]/[HA]) + \log \gamma_\pm \quad (4.29)$$

In eq. (4.29), $[A^-]$ and $[HA]$ refer to the *free* concentrations of these species in solution. The use of *stoichiometric* concentrations of the acid and base forms in eq. (4.29) is valid only where association equilibria, such as homohydrogen-bond formation, are negligible, except in the particular case of $[A^-] = [HA]$, i.e., the half-neutralization point. More generally, the relationship between pH, K_a, K_{AHA} and the stoichiometric concentrations of acid and base during a titration is given in Appendix 4.3 (eq. A4.12). It is also important in lower-dielectric solvents to take into account ion-pair formation between A^- and the accompanying cation, M^+, in determining the free concentration of $[A^-]$ for use in eq. (4.29) (Appendix 4.2).

In practice it is often convenient to calibrate the glass electrode in standard *aqueous* buffer solutions and then use it directly to monitor the titration of the unknown acid in the non-aqueous or mixed-aqueous solvent. The 'apparent' pH so measured, pH^{ap}, will differ from the true pH, pH^S, by an amount δ, according to eq. (4.30), as will the 'apparent' pK_a derived from the titration. The difference in pH-scales is illustrated in Fig. 4.3.

$$pH^{ap} = pH^S + \delta \quad (4.30)$$

The correction, δ, is determined by measuring pH^{ap} in a solution of known pH, pH^S, in the solvent in question, or by repeating the titration using an acid of know pK_a in the solvent.

In mixed aqueous–organic solvents, pH-values of standard buffers are often available [1, 16]. Alternatively, strong acids such as HCl, methanesulfonic

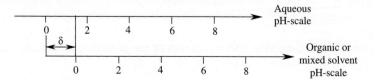

Fig. 4.3.
The relationship between pH in aqueous and non-aqueous or mixed-aqueous solvents

Experimental methods for the determination of dissociation constants

acid, etc., suffice to provide calibration solutions of known pH, via the stoichiometric acid concentration.

Grunwald and co-workers [17] describe a very convenient method for determining the dissociation constants of weak acids, such as carboxylic acids, which is generally applicable to aqueous–organic mixtures, and which includes an electrode calibration as part of the titration. It is based on cell 4.31, which contains a mixture of a strong acid, e.g., HCl, and the acid, HA, whose pK_a is to be determined.

$$\text{Glass electrode}|H^+, Cl^-, HA, \text{methanol} - \text{water}|AgCl, Ag \qquad (4.31)$$

A known amount of strong acid (HCl) is added to the solution of the weak acid, HA, and the EMF (pH^{ap}) is measured. Sufficient NaOH to neutralize all of the HCl, and some of the HA is then added. In the initial solution, the HCl effectively suppresses the dissociation of HA, and thus the hydrogen ion concentration equals [HCl]; the measured EMF (E^R) then serves to calibrate the system (eq. (4.27)). The second solution contains a known ratio of $[A^-]/[HA]$ (from partial neutralization of HA by the excess NaOH), and the EMF measurement (E_X, eq. (4.28). gives the pH of the solution; this can be used in conjunction with eq. (4.29) to determine the pK_a of HA. An important feature of the cell is that it has no liquid junction. The method is generally applicable in aqueous–organic mixtures.

The method is illustrated in Fig. 4.4, which shows the change in measured pH, pH^{ap}, during the titration of 40 ml of HCl (0.01M) and acetic acid, HOAc (0.0057 M), with NaOH (0.2 M), all in MeOH–water (60 vol%, 54 wt%) at 20°C [18].

Two endpoints are observed—the first corresponding to the neutralization of the HCl and the second to neutralization of HOAc. Points prior to the first endpoint, region (a), comprise solutions of known $[H^+]$ ($[H^+] = [HCl]_T - [NaOH]$, where $[HCl]_T$ is the total concentration of HCl, and [NaOH] is the concentration of added NaOH); hence they can be used to calculate the true pH and, by comparison with pH^{ap}, the correction factor, δ, eq. (4.30). Points lying between the two endpoints, region (b), refer to known ratios of $[OAc^-]/[HOAc]$ and can thus be used in conjunction with the corrected pH to determine the pK_a of HOAc in the solvent mixture, as described above.

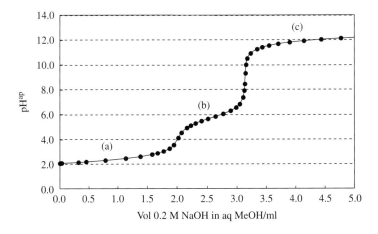

Fig. 4.4.
pH-titration of HCl and acetic acid with NaOH in 54 wt% methanol–water

*Points in region (c) correspond to pH-values measured in solutions of known total base concentration and can be used to determine the autoionization constant of the mixed solvent, Sections 4.6 and 5.1

For any point in this region, $[OAc^-] = [HOAc]_T + [HCl]_T - [NaOH]$, and $[HOAc] = [HOAc]_T - [OAc^-]$.*

The method is equally applicable to the determination of pK_a values of cationic acids, such as Et_3NH^+, illustrated in Fig. 4.5, which shows the change in measured pH, pH^{ap}, during the titration of 40 ml of Et_3N (0.01M) with HCl (0.2 M), all in MeOH–water (60 vol%, 54 wt%) at 20°C [18]. In this case though, the electrode is calibrated using the data points beyond the endpoint, i.e., region (b), which correspond to known [HCl]. Points lying in region (a), which correspond to known ratios of $[Et_3N]/[Et_3NH^+]$, can then be used to determine the pK_a of Et_3NH^+ in the solvent mixture.

It is normally not possible to use such a simple cell in polar aprotic solvents, because of the problem of finding a suitable reference electrode. Thus, for example, the Ag,AgCl reference electrode cannot be used because the high activity of the chloride ion in aprotic solvents leads to solubilization of AgCl by formation of $AgCl_2^-$ and higher chloro-complexes [19]. In much of the work described by Kolthoff and co-workers, the reference electrode used was Ag, Ag^+, which is connected via a salt-bridge to the hydrogen or glass electrode [8–10, 14, 15].

4.5.2 Acid–base indicators

The use of acid–base indicators, HIn, in determining pK_a-values is long established in aqueous solution, and the methodology is equally applicable to non-aqueous media. In pure non-aqueous media it is generally more convenient for the non-specialist than the potentiometric method, and most recent comprehensive data compilations have involved the use of spectrophotometric methods. We will describe the procedure using as an example indicators from the extensive work by Bordwell and co-workers in dimethylsulphoxide (DMSO) and N-methylpyrrolidin-2-one (NMP) as solvents [6, 20, 21].

Representative indicators are given in Scheme 4.2, together with their pK_a-values in DMSO.

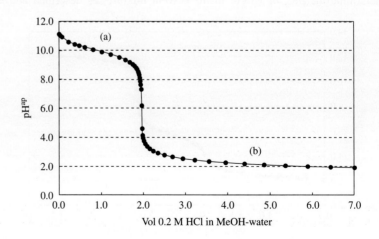

Fig. 4.5.
pH-titration of Et_3N HCl in 54 wt% methanol–water

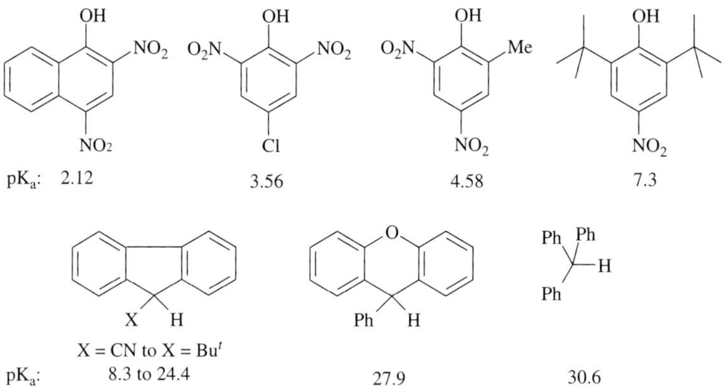

Experimental methods for the determination of dissociation constants

pK$_a$: 2.12, 3.56, 4.58, 7.3

pK$_a$: X = CN to X = Bu^t, 8.3 to 24.4; 27.9; 30.6

Scheme 4.2.
Acid-base indicators in dimethylsulfoxide

The first stage consists of determining the pK$_a$ of the indicators, typically beginning with the most acidic, in the 0-6 pK$_a$-region. The equilibria is given by eq. (4.32), in which α represents the degree of ionization.

$$\text{HIn} \underset{1-\alpha}{\overset{K_a}{\rightleftharpoons}} \underset{\alpha}{H^+} + \underset{\alpha}{In^-} \quad (4.32)$$

(i) The molar absorption coefficient of the anion, ε_{In}, is determined at a fixed wavelength by adding an excess of Et$_3$N to a known amount of HIn and measuring the absorbance of the anion. (ii) Aliquots of HIn are then added to the pure solvent, and the absorption of the anion generated by autoionization, eq. (4.32), is recorded. The concentration of the anion and hence the degree of ionization is calculated from ε_{In}; then, $K_a = \alpha^2/(1-\alpha)$.

The next step is the construction of a ladder anchored on the more highly acidic indicators, or other acids, such as malonitrile (pK$_a$ = 11.1), whose pK$_a$-values have been accurately determined by the potentiometric method. The procedure is illustrated in Scheme 4.3, using as an example 9-(phenylthio)fluorene (pK$_a$ = 15.4) and 9-(phenyl)fluorine (pK$_a$ = 17.9), which are linked via 1,1-diethylsulphonylethane (pK$_a$ = 17.0).

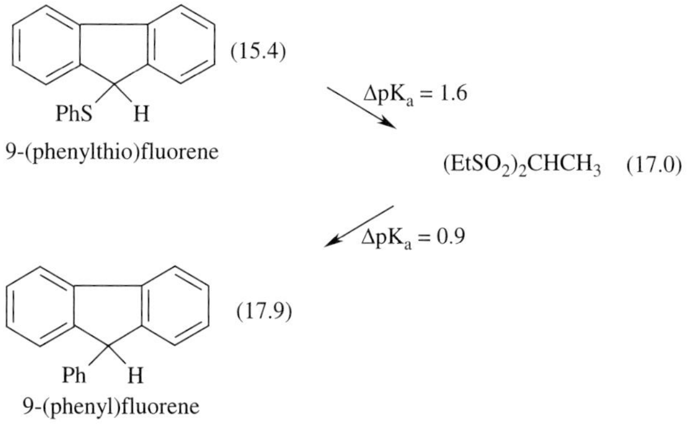

9-(phenylthio)fluorene (15.4)
ΔpK$_a$ = 1.6
(EtSO$_2$)$_2$CHCH$_3$ (17.0)
ΔpK$_a$ = 0.9
9-(phenyl)fluorene (17.9)

Scheme 4.3.
Ladder construction for 9-(phenylthio)fluorene and 9-(phenyl)fluorene in dimethylsulfoxide

In the first stage, the ratio of 9-(phenylthio)fluorene (HIn) to its anion, In$^-$, in the presence of a known ratio of (EtSO$_2$)$_2$CHCH$_3$ (HA) and its anion, A⁻, is measured spectrophotometrically. The ratios can then be substituted in eq. (4.33), which links the two pK_a-values, to give the pK_a of $(EtSO_2)_2CHCH_3$.

$$pK_a(HA) = pK_a(HIn) + \log \frac{[HA][In^-]}{[A^-][HIn]} \tag{4.33}$$

The unknown 9-(phenyl)fluorene can then be equilibrated with fixed solution of $(EtSO_2)_2CHCH_3$ and its anion, and the ratio of its acid to anion forms measured spectrophotometrically. Substitution of the appropriate values in eq. (4.33), in which HIn and In⁻ now represent 9-(phenyl)fluorene and its anion, then gives the desired pK_a-value.

In this manner, indicators covering a pK_a range from 2.1–30.6 in DMSO, and similarly in NMP (2.7–31), have been established and used to determine pK_a-values for a wide range of substrates.

Solutions of known ratios of HA/A⁻ for the more weakly acidic substrates in DMSO are typically generated by addition of the strongly basic potassium or caesium dimsyl salts, $[M^+CH_3(SO)CH_2^-]$, to AH. The dimsyl salts are prepared either by the addition of KH or $CsNH_2$ (followed by degassing to remove NH_3) to DMSO, and similarly the conjugate base of NMP, $[K^+C_5H_8NO^-]$.

A comprehensive but structurally different set of indicators has also been reported for CH_3CN, covering a range of 28 pK-units, and including some 89 bases [22]. Relative pK_a-values of successive bases, eq. (4.34), were determined spectrophotometrically, and the scale was anchored on the value for pyridine, $pK_a = 12.53$, derived from a large number of independent spectrophotometric and potentiometric measurements.

$$B_1H^+ + B_2 \rightleftharpoons B_1H^+ + B_2$$

$$pK_a(B_2H^+) = pK_a(B_1H^+) + \log \frac{[B_1][B_2H^+]}{[B_1H^+][B_2]} \tag{4.34}$$

Representative bases, many of which are substituted pyridines or anilines, are given in Scheme 4.4, together with their pK_a-values. The strongest bases are the so-called 'P_n' bases, as in $PhP_{1-3}(dma)$, introduced by Schwesinger [23].

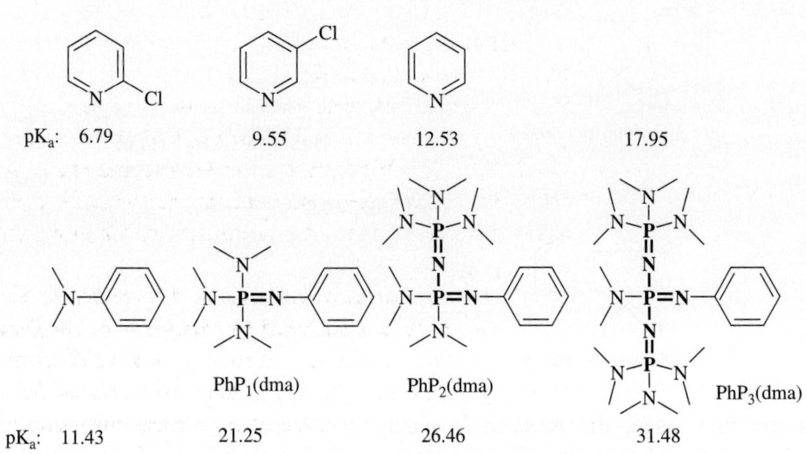

pK_a: 6.79   9.55   12.53   17.95

         PhP₁(dma)   PhP₂(dma)   PhP₃(dma)

pK_a: 11.43   21.25   26.46   31.48

Scheme 4.4.
Bases in acetonitrile

The use of these bases to determine the pK_a's of a series of calixarenes, such as calix[4]arene, in acetonitrile has been described by Cunningham [24].

4.6 Autoionization constants of solvents*

*Also known as autoprotolysis constants or ionic products

An important property of any solvent is the tendency towards self ionization, represented by eq. (4.35) for solvent SH.

$$SH + SH \xrightleftharpoons{K_{ai}} SH_2^+ + S^- \qquad (4.35)$$

In eq. (4.35), K_{ai} is the autoionization constant, which provides a quantitative measure of this self-ionization. In a pure solvent, SH, the definition of the autoionization constant is straightforward. Following our convention in Chapter 2, in which the symbol H^+ represents the solvated proton, and the (constant) solvent concentration is included within the equilibrium constant, the autoionization constant, K_{ai}, is given by eq. (4.36).

$$K_{ai} = [H^+][S^-]\gamma_\pm^2 \qquad (4.36)$$

In eq. (4.36), $\gamma_\pm$ is the mean activity coefficient of H^+ and S^- (eq. (4.16)), with respect to infinite dilution in solvent SH.

Autoionization constants provide a ready indication of the maximum available pH in a particular solvent. Thus, for a concentration of $[S^-] = 1$ M in a solvent with autoionization constant, K_{IP}, it follows from eq. (4.36) that pH $\sim$ pK_{ai}, where p$K_{ai} = -\log K_{ai}$; for example pH $\sim$ 14 in water ($K_{ai} = K_w = 10^{-14}$ M^2) and $\sim$ 28.5 in t-BuOH ($K_{ai} = 10^{-28.5}$ M^2, Table 5.1). The magnitude of the pH-jump at the endpoint for acid–base titrations also depends directly on the value of the autoionization constant [1, 25], and this is often used to advantage in analytical methods based upon pH-titrations in partially or wholly non-aqueous media.

The most convenient and widely used method of determining K_{ai} values is by EMF measurements on cells of the type shown in eq. (4.37), illustrated for the case of MeOH; the hydrogen electrode can be, and is most commonly, replaced by a glass electrode.

$$\text{Pt, H}_2(g) \mid \text{NaOMe, KCl in MeOH} \mid \text{AgCl, Ag} \qquad (4.37)$$

The cell reactions and the cell potential, E, are given by eqs. (4.38) and (4.39), respectively, where E^o is the standard cell potential[†].

[†]Activity coefficients have been omitted for clarity; they may be calculated using the Davies equation (4.19) or eliminated by extrapolating measurements at different concentrations to infinite dilution

$$\text{AgCl} + e \rightarrow \text{Ag} + \text{Cl}^- \qquad (4.38a)$$

$$H^+ + e \rightarrow \tfrac{1}{2}H_2(g) \qquad (4.38b)$$

$$E = E^o - RT/F \ln[H^+][Cl^-] \qquad (4.39)$$

The hydrogen-ion concentration in cell (4.37) is, however, controlled by the autoionization equilibrium, eq. (4.36), from which we obtain $[H^+] = K_{ai}/[\text{MeO}^-]$. Substituting into eq. (4.39) and rearranging gives eq. (4.40), where p$K_{ai} = -\log K_{ai}$, and the factor 2.303 results from conversion from natural logarithms to $\log_{10}$.

$$(E - E^o)F/2.303RT = pK_{ai} + \log[\text{MeO}^-]/[\text{Cl}^-] \qquad (4.40)$$

K_{ai} can then be determined by measurement of the cell potential (cell(4.37)) at various known concentrations of methoxide and chloride and substituting into eq. (4.40), provided that E^o is known. The required standard cell potential, E^o, can in turn be readily determined by measurement of the cell potential in the presence of known concentrations of HCl, cell 4.41, and substituting into eq. (4.39).

$$\text{P}_t, \text{H}_2(\text{g})|\text{HCl in MeOH}|\text{AgCl, Ag} \qquad (4.41)$$

Appendix 4.1 Dissociation of acetic acid in the presence of sodium chloride [26]

Table 4A.1 shows the dependence of the dissociation constant for acetic acid in the presence in increasing levels of added sodium chloride.

The pK_a expressed in terms of solution concentrations is lower by some 0.2 units in 0.33M NaCl compared with dilute solutions, and this decrease can be essentially entirely accounted for by the activity coefficients of H^+ and OAc^-, using eqs. (4.17) and (4.19), as shown by the figures in final column of Table 4A.1. Corresponding effects in non-aqueous media will be larger, because of the stronger sensitivity to ionic strength (Table 4.1).

At very high concentrations of NaCl (e.g., > 1M) the observed pK_a-values begin to increase with increasing salt concentration, largely because of the reduction in water activity which reduces its effectiveness in solvating the ions.

Appendix 4.2 Ion-pair formation and pK_a-determination

For acid HA in lower dielectric solvents, the anion, A^-, will typically be involved ion-pair equilibria with the corresponding cation, M^+, eq. (A4.1).

Table 4A.1 The dissociation constant of acetic acid in aqueous sodium chloride at 25°C

$I = [\text{NaCl}]/\text{M}$	pK_a^a	$-\log \gamma_\pm^b$	pK_a^{oc}
0.0044	4.695	0.039	4.757
0.0128	4.658	0.050	4.758
0.0229	4.632	0.064	4.759
0.0397	4.606	0.079	4.763
0.0529	4.589	0.087	4.764
0.1278	4.541	0.115	4.771
0.3327	4.489	0.136	4.760

[a] $K_a = [\text{H}^+][\text{OAc}^-]/[\text{HOAc}]$; [b] Eq. (4.19); [c] $pK_a^o = pK_a - 2\log \gamma_\pm$, eq. (4.17)

$$M^+ + A^- \underset{}{\overset{K_{IP}}{\rightleftharpoons}} (M^+A^-) \qquad (A4.1)$$
$$(a-x) \quad (a-x) \qquad\qquad x$$

In eq. (4.11), a represents the stoichiometric concentration of anion A^- (and cation M^+), and the ion-pair formation constant, K_{IP}, is given by eq. (A4.2).

$$K_{IP} = \frac{x}{(a-x)^2} \qquad (A4.2)$$

This can be solved as a simple quadratic to determine x and hence the free concentration of $A^- (= a - x)$ to be used in eq. (4.29).

K_{IP}-values are typically determined from conductimetric measurements on solutions containing different concentrations of (M^+A^-) [4–7], or by monitoring the effect of added salts on indicator absorbance [5].

Appendix 4.3 Determination of homohydrogen-bond association constants, K_{AHA}

The coupled equilibria involving acid dissociation and homohydrogen-bond formation for acid HA, such as a carboxylic acid, are given in eq. (A4.3).

$$RCO_2H \overset{K_a}{\rightleftharpoons} H^+ + RCO_2^-$$

$$RCO_2H + RCO_2^- \overset{K_{AHA}}{\rightleftharpoons} RCO_2^- \cdots HO_2CR \qquad (A4.3)$$

There are two approaches to the measurement of K_{AHA}: direct determination of the interaction of HA and A^-, through either solubility or spectrophotometric measurements; and indirect determination via the influence of the association on the pH of mixtures of HA and A^-.

A4.3.1 Direct measurement of K_{AHA}

Solubility measurements

Simple salts of carboxylic acids or phenols often have only limited solubility in typical aprotic solvents. Addition of the conjugate acid then causes an increase in solubility as a result of homohydrogen-bond formation between the acid and the anion. For example, sodium methanesulfonate has a solubility of 1.2×10^{-4}M in acetonitrile at 25°C, and this increases more than ten-fold to 3.0×10^{-3}M on addition of 5.0×10^{-2}M methanesulphonic acid.

Determination of K_{AHA} from the increase in solubility is straightforward. Thus, the dissolution of sodium methanesulphonate, $[Na^+A^-]$, in the presence of methanesulphonic acid, HA, may be represented by eq. (A4.4).

$$[Na^+A^-]_c \underset{}{\overset{K_{sp}}{\rightleftharpoons}} Na^+ + A^-$$

$$A^- + HA \underset{}{\overset{K_{AHA}}{\rightleftharpoons}} AHA^- \qquad (A4.4)$$

The various equilibria and mass- and charge-balance equations linking the species are given in eqs. (A4.5)–(A4.8), in which C_{HA} is the total amount of HA added and $[Na^+] = C_A$ is the amount of the salt in solution.

$$K_{sp} = [Na^+][A^-] \qquad (A4.5)$$

$$K_{AHA} = \frac{[AHA^-]}{[A^-][HA]} \qquad (A4.6)$$

$$C_A = [Na^+] = [A^-] + [AHA^-] \qquad (A4.7)$$

$$C_{HA} = [HA] + [AHA^-] \qquad (A4.8)$$

Substituting $[A^-] = K_{sp}/[Na^+]$ from eq. (A4.5) into eq. (A4.7), and the resultant expression for $[AHA^-] = [Na^+] - K_{sp}/[Na^+]$ into eq. (A4.8), gives all three species in eq. (A4.4) in terms of known quantities, and hence eq. (A4.9).

$$K_{AHA} = \frac{([Na^+] - K_{SP}/[Na^+])}{(K_{SP}/[Na^+])(C_{HA} - \{[Na^+] - K_{SP}/[Na^+]\})} \qquad (A4.9)$$

The measurements required for the calculation of K_{AHA} from eq. (A4.9) are the solubility of NaA in the absence of added HA, hence K_{sp}, and the total solubility, $C_A = [Na^+]$, in the presence of known amounts of added HA.

Spectrophotometric measurements

The characteristic UV/Vis absorption bands of, in particular, phenoxide ions show a spectral shift as a result of homohydrogen-bond formation. The 3,5-dinitrophenoxide ion exhibits a striking visual effect upon homohydrogen-bond formation; the simple ion is red in acetonitrile, while the homohydrogen-bonded ion is yellow. In a typical procedure the absorbance of a soluble tetra-alkyl ammonium phenoxide at an appropriate wavelength is measured in presence of known amounts of phenol. The total absorbance, A, is given by eq. (A4.10), in which ε_A and ε_{AHA} are the absorption coefficients of A^- and AHA^-, respectively.

$$A = \varepsilon_A[A^-] + \varepsilon_{AHA}[AHA^-] \qquad (A4.10)$$

Substituting $[A^-] = (A - \varepsilon_{AHA}[AHA^-])/\varepsilon_A$ into $C_A = [A^-] + [AHA^-]$ (eq. (A4.7)), and subsequently for $[AHA^-]$ into eq. (A4.10), gives expressions for $[A^-]$, $[AHA^-]$ and $[HA]$ in terms of ε_{AHA} and the known ε_A, A, $[A^-]_t$ and C_{HA} for substitution into eq. (A4.4). The resultant expression, eq. (A4.11), in which $[AHA^-] = (C_{HA} - A/\varepsilon_A)/(1 - \varepsilon_{AHA}/\varepsilon_A)$, can be fitted for various values of A, $[A^-]_t$, and $[HA]_t$ to give both K_{AHA} and ε_{AHA}, using EXCEL or any number of standard data-fitting packages.

$$K_{AHA} = \frac{[AHA^-]}{(A - \varepsilon_{AHA}[AHA^-]/\varepsilon_A)(C_{HA} - [AHA^-])} \qquad (A4.11)$$

Alternatively, if the association constant is sufficiently large, ε_{AHA} can be measured directly by using a suitable excess of HA.

A4.3.2 Indirect measurement of K_{AHA}

Influence of K_{AHA} on pH

It was shown in Section 4.1.5 (Fig. 4.1) that a pH-titration curve in the presence of homohydrogen-bonding deviates increasingly from that for the simple case ($K_{HAA} = 0$), as the distance from the point of half-neutralisation, $(pH)_{½} = pK_a$, increases. It is possible in a straightforward manner, by combining $K_a = [H^+][A^-]/[HA]$ with eqs. (A4.7) and (A4.8), to derive a relationship between K_{AHA} and the solution pH for various values of total acid, C_{HA}, and anion, C_A, during the titration.* This is given by eq. (A4.12), in which $r = [H^+]/K_{HA}$. Kolthoff, Chantooni, and Bhowmik [9] give full details.

$$K_{AHA} = \frac{r^2 C_A - r(C_{HA} + C_A) + C_{HA}}{r(C_A - C_{HA})^2} \quad (A4.12)$$

Eq. (A4.12) can be fitted for various H^+, C_{HA}, and C_A to derive both K_{AHA} and K_a. Alternatively, K_a can be obtained from the half-neutralization point of the titration and used in conjunction with $r = [H^+]/K_a$ to calculate K_{AHA}.

*Thus, $[HA] = [A^-]([H^+]/K_a) = r[A^-]$, $[AHA^-] = C_A - [A^-]$ (eq. (A4.7)). Substitution into eq. (A4.8) gives $[A^-] = (C_A - C_{HA})/(1 - r)$; hence $[HA] = r(C_A - C_{HA})/(1 - r)$ and $[AHA^-] = (C_{HA} - rC_A)/(1 - r)$. Substitute for A^-, HA, and AHA^- into eq. (A4.6)

References

[1] Bates, R. G. 'Determination of pH: Theory and Practice', Wiley-Interscience, N.Y., 1973
[2] Bates, R. G., Guggenheim, E. A. *Pure & Appl. Chem.* 1960, *1*, 163
[3] Davies, C. W. 'Ion Association', Butterworths, London, 1962
[4] Barrón, D., Barbosa, J. *Anal. Chim. Acta*, 2000, *403*, 339
[5] Olmstead, W. N., Bordwell, F. G. *J. Org. Chem.*, 1980, *45*, 3299
[6] Bordwell, F. G., Branca, J. C., Hughes, D. L., Olmstead, W. N. *J. Org. Chem.* 1980, *45*, 3305
[7] Juillard, J., Kolthoff, I. M. *J. Phys. Chem.*, 1971, *75*, 2496
[8] Izutsu, K., Kolthoff, I. M., Fujinaga, Y., Hattori, M., Chantooni, M. K. *Anal. Chem.*, 1977, *49*, 503
[9] Kolthoff, I. M., Chantooni, M. K., Bhowmik, S. *J. Am. Chem. Soc.*, 1966, *88*, 5430
[10] Coetzee, J. F., Padmanabhan, G. R. *J. Phys. Chem.*, 1965, *69*, 3193
[11] Kolthoff, I. M., Chantooni, M. K., Bhowmik, S. *J. Am. Chem. Soc.*, 1968, *90*, 23
[12] Augustin-Nowacka, D., Makowski, M., Chmurzynski, L. *Anal. Chim. Acta,* 2000, *418*, 233
[13] Coetzee, J. F., Padmanabhan, G. R. *J. Am. Chem. Soc.*, 1965, *87*, 5005
[14] Kolthoff, I. M., Chantooni, M. K. *J. Am. Chem. Soc.*, 1965, *87*, 4428
[15] Kolthoff, I. M., Chantooni, M. K. *J. Am. Chem. Soc.*, 1971, *93*. 3843
[16] Mussini, T. Covington, A. K., Dal Pozzo, F., Longhi, P., Rondinini, S., Zou, Z.-Y. *Electrochim. Acta*, 1983, *28*, 1593
[17] Bacarella, A. L., Grunwald, E., Marshall, H. P., Purlee, E. L. *J. Org. Chem.*, 1955, *20*, 747
[18] Chan, L. C. Personal communication
[19] Coetzee, J. F., Deshmukh, B. K., Liao, C.-C. *Chem. Rev.*, 1990, *90*, 827
[20] Matthews, W. S., Bares, J. E., Bartmess, J. E., Bordwell, F. G., Cornforth, F. J., Drucker, G. E., Margolin, Z. McCallum, R. J., McCollom, G. J., Vanier, N. R. *J. Am. Chem. Soc.,* 1975, *97*, 7006
[21] Bordwell, F. G. *Acc. Chem. Res.*, 1988, *21*, 456

[22] Kaljurand, I., Kütt, A., Sooväli, L., Rodima, T., Mäemets, V., Leito, I., Koppel, I. A. *J. Org. Chem.*, 2005, *70*, 1019
[23] Schwesinger, R., Schlemper, H., Hasenfratz, C., Willaredt, J., Dambacher, T., Breuer, T., Ottaway, C., Fletschinger, M., Boele, J., Fritz, H., Putzas, D., Rotter, H. W., Bordwell, F. G., Satish, A. V. Ji, G.-Z., Peters. E.-M., Peters, K., von Schnering, H. G., Walz, L. *Liebigs Ann.*, 1996, 1055
[24] Cunningham, I. D., Woolfall, M. *J. Org. Chem.*, 2005, *70*, 9248
[25] Rondinini, S., Longhi, P., Mussini, P. R., Mussini, T. *Pure & Appl. Chem.*, 1987, *59*, 1693
[26] Kilpi, S., Lindell, E. *Suomen Kemi.*, 1963, *36*, 81

Protic Solvents

5

Protic solvents include water, alcohols, formamide and other primary and secondary amides, and formic acid. A defining characteristic is their ability to form strong hydrogen bonds with suitable acceptors, which arises from the fact that they have hydrogen bound directly to electronegative atoms, such as oxygen and nitrogen.*

*See, however, liquid NH_3, Chapter 6.

Alcohols and alcohol–water mixtures in particular are widely used as solvents in synthetic, analytical and crystallization processes. They are better at dissolving organic molecules than water, have a wider pH range, and are able to support reasonable concentrations of ionic species.

5.1 Autoionization constants

Autoionization constants, $K_{ai} = [H^+][RO^-]\gamma_{\pm}^2$ (Section 4.6) [1, 2], have been determined for a variety of alcohols, and are reported in Table 5.1.

All alcohols show values higher than that for water ($pK_{ai} = pK_w = 14.00$ at 25°C) and can therefore support stronger bases than water, especially t-BuOH.

Autoionization equilibria are more complex in alcohol–water mixtures than in the pure solvents because of the coexistence of hydroxide and alkoxide ions, e.g., eq. (5.1) for methanol–water [1, 3, 4].

$$H_2O \rightleftharpoons H^+ + OH^-$$

$$MeOH \rightleftharpoons H^+ + MeO^-$$

$$OH^- + MeOH \rightleftharpoons MeO^- + H_2O \quad (5.1)$$

We may, however, define and measure an apparent ionic product, K_{ai}, in the solvent mixtures, given by eq. (5.2), in which K_{ai}^W and K_{ai}^M are the autoionization constants of water and methanol, respectively, *in the mixtures*.

$$K_{ai} = [H^+]\{[OH^-] + [MeO^-]\} = K_{ai}^W + K_{ai}^M \quad (5.2)$$

Measurement of K_{ai}-values in mixed-aqueous solvents is straightforward. For example, in methanol–water, a simple titration of HCl with NaOH, analogous to that shown in Fig. 4.4 (Chapter 4), but not including the acetic acid, allows calibration of the electrode from pH values measured prior to the endpoint

Table 5.1 Autoionisation constants (pK_{ai}) of protic solvents at 25°C[a]

Solvent	pK_{ai}	Solvent	pK_{ai}
MeOH	16.71	t-BuOH	28.5
EtOH	18.80	n-pentanol	20.65
n-PrOH	19.30	n-hexanol	19.74
i-PrOH	20.80	formamide	16.8
n-BuOH	21.56	formic acid	6.5

[a] Ref [1]

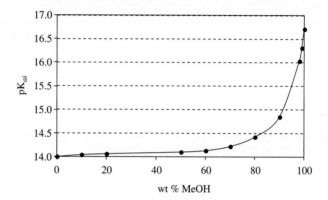

Fig. 5.1. Autoionisation constants, pK_{ai}, in methanol-water mixtures [1, 3, 4]

(region (a) in Fig. 4.4) and hence the pH([H$^+$]) in solutions of known total base ([OH$^-$] + [MeO$^-$]) beyond the endpoint (as in region (c) in Fig. 4.4), for substitution into eq. (5.2).

Autoionization constants have been reported in both methanol–water [1, 3, 4] and ethanol–water [5] mixtures and show similar variations with solvent composition. In particular, the major changes in pK_{ai} occur at high alcohol content of the solvent, from about 80 wt% onwards. This is shown in Fig. 5.1 for methanol–water mixtures: there is little change up to 50 wt%, from which point there is an increasingly rapid change with composition to the final value in pure methanol of $pK_{ai}^M = 16.71$. The ions in the mixtures are clearly preferentially solvated by water (Chapter 3, Section 3.3.2).

Calculations based upon the free energies of transfer of the ions among the solvent mixtures suggest that the relative fractions of hydroxide and methoxide in the mixtures closely parallel the relative solvent fractions, e.g., in 80 wt% MeOH, methoxide ions comprise around 82% of the total anion concentration [3, 6].

5.2 Methanol

Methanol has a relatively low dielectric constant, $\varepsilon_r = 32.6$ (Table 1.1), and hence electrostatic interactions between ions are significant even at relatively low concentrations; e.g., at a total concentration (ionic strength) of 0.01M,

$\gamma_\pm = 0.68$, $\log \gamma_\pm = -0.17$ (Section 4.1.3). The result is that dissociation constants for carboxylic acids and phenols expressed in terms of simple concentration quotients are very sensitive to the solution ionic strength.

Reference pH-values for electrode calibration in methanol–water and pure methanol have been reported by Bates [7]. These are based on oxalate (oxalic acid and ammonium hydrogen oxalate, each at 0.01 mol kg^{-1}) and succinate (succinic acid and lithium hydrogen succinate, each at 0.01 mol kg^{-1}) buffers, which have pH-values of 5.79 and 8.75, respectively, in methanol at 25°C.

The most extensive compilation of dissociation constants in methanol has been reported by Bosch and co-workers [8]. It includes aliphatic and aromatic carboxylic acids, phenols, and a variety of amines and related nitrogen bases. The results all refer to infinite dilution in the solvent and they can be corrected as required to allow for the influence of higher ionic strengths by use of the Davies equation (4.19). Representative values are discussed below, and more extensive tabulations of data in methanol are included in Appendix 9.1.

5.2.1 Neutral acids: carboxylic acids, phenols

Carboxylic acids

Dissociation constants for a variety of aliphatic and aromatic carboxylic acids are listed in Table 5.2. Aqueous values are included for comparison.

There is a good correlation between the pK_a-values in methanol and in water, as illustrated in Fig. 5.2 for the aliphatic carboxylic acids.

Altogether some 117 carboxylic acids have been measured and the data, including both aromatic and aliphatic acids, can be represented by eq. (5.3) [8].

$$\text{p}K_a(\text{MeOH}) = 1.02\,\text{p}K_a(\text{H}_2\text{O}) + 4.98 \qquad (5.3)$$

Table 5.2 pK_a-values of carboxylic acids in methanol at 25°C^a

Aliphatic acid	pK_a MeOH	pK_a H$_2$O	ΔpK_a	Benzoic acid	pK_a MeOH	pK_a H$_2$O	ΔpK_a
2,2-dichloroacetic	6.38	1.34	5.04	2,6-dichloro	7.05	1.82	5.23
2-cyanoacetic	7.50	2.46	5.04	2-nitro	7.64	2.19	5.45
2-fluoroacetic	7.99	2.82	5.17	3,5-dinitro	7.38	2.67	4.71
2-chloroacetic	7.88	2.85	5.03	2-chloro	8.31	2.92	5.39
2-chloropropanoic	8.06	2.90	5.16	4-nitro	8.34	3.43	4.91
2-bromoacetic	8.06	2.90	5.16	3-nitro	8.32	3.47	4.85
2-bromopropanoic	8.22	3.00	5.22	3-chloro	8.83	3.80	5.03
2-iodoacetic	8.36	3.13	5.23	3-bromo	8.80	3.83	4.97
2-hydroxyacetic	8.68	3.85	4.83	4-cyano	8.42	3.53	4.89
3-bromopropanoic	9.00	4.04	4.96	3-cyano	8.53	3.60	4.93
acetic	9.72	4.75	4.97	4-bromo	8.93	3.99	4.94
butanoic	9.69	4.82	4.87	3-methoxy	9.30	4.12	5.18
propanoic	9.71	4.88	4.83	4-chloro	9.09	4.00	5.09
				H	9.38	4.19	5.19
Benzoic acid:				3-methyl	9.39	4.28	5.11
2,6-dinitro	6.30	1.14	5.16	4-methyl	9.51	4.38	5.13
2,4-dinitro	6.45	1.43	5.02	4-hydroxy	9.99	4.55	5.44

^{a} Data from ref [8]

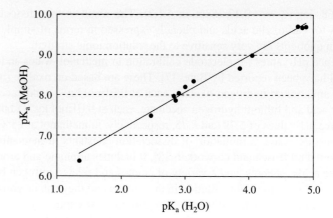

Fig. 5.2.
pK$_a$-values for aliphatic carboxylic acids in methanol versus water at 25°C

Eq. (5.3) shows that there is an almost constant increase of 5.0 pK units on transferring a carboxylic acid from water to methanol. It follows, therefore, that substituent effects on the dissociation constants of carboxylic acids are essentially the same in methanol as in water.

The pK$_a$ difference between water and methanol may be analysed in terms of the free energy change for the various species, eq. (5.4), eq. (3.4).

$$\mathrm{pK_a(MeOH)} - \mathrm{pK_a(H_2O)} = \{\Delta G_{tr}(H^+) + \Delta G(A^-) \\ - \Delta G_{tr}(HA)\}/2.303RT \quad (5.4)$$

Three factors contribute to the increase of five units in the pK$_a$ of the acids in methanol relative to water: modest increases in the free energies of the proton and carboxylate groups, and a decrease in that of the carboxylic acid [9]. The poorer solvation of the proton in methanol (Table 3.5) equates to an increase in pK$_a$ of $\sim$ 2 units. The remainder, $\sim$ 3 units (17 kJ mol^{-1}), is independent of the substrate, and therefore represents essentially the different responses of the carboxylate and carboxylic groups to the transfer between water and methanol.

Aliphatic dicarboxylic acids, including especially maleic, fumaric, and succinic acids, are frequently used in the formulation of pharmaceutical actives, through salt formation with nitrogen bases, and their dissociation constants are consequently of some interest. Representative acids are included in Table 5.3, along with the dissociation constants for the first and second dissociations [10].

pK$_{a1}$-values, corresponding to the first dissociation of the acids to form the monoanions, mostly behave very similarly to those of the monocarboxylic acids (Table 5.2), showing an average increase of around five units on transfer from water. Maleic acid and, to a lesser extent, malonic acid exhibit noticeably smaller increases, which may be attributed to a modest stabilization of the monoanion by intramolecular H-bonding, thereby increasing the tendency to dissociate and hence reducing pK$_{a1}$.

pK$_{a2}$-values show slightly larger increases, on average around 6.2 units, indicating that formation of the second carboxylate anion is more sensitive to

Table 5.3 pK_a-values of dicarboxylic acids in methanol at 25°C[a]

Dicarboxylic acid	pKa_1 MeOH	$pK_{a1}H_2O$	ΔpK_{a1}[b]	pK_{a2}MeOH	$pK_{a2}H_2O$	ΔpK_{a2}[b]
oxalic	6.10	1.23	4.87	10.7	4.29	6.4
maleic[c]	5.7	1.92	3.8	12.8	6.24	6.6
malonic	7.50	2.87	4.63	12.4	5.67	6.7
fumaric[c]	7.9	3.02	4.9	10.3	4.38	5.9
succinic	9.10	4.20	4.90	11.5	5.55	6.0
glutamic	9.40	4.34	5.06	11.5	5.42	6.1
adipic	9.45	4.43	5.02	11.1	5.42	5.7

[a] Ref [10]; [b] $\Delta pK_a = pK_a(MeOH) - pK_a(H_2O)$; [c] Kolthoff, I. M., Chantooni, M. K. *Anal. Chem.*, 1978, *50*, 1440; Garrido, G., de Nogales, V., Ràfols, C. Bosch, E. *Talanta*, 2007, *73*, 115

solvent change than the first. Presumably with a sufficiently large separation, the two carboxylate groups would act independently, but even for adipic acid ($HO_2C(CH_2)_4CO_2H$) the increase in pK_a for the second carboxylate group remains at 5.7 units.

Phenols

Table 5.4 lists the dissociation constants for phenols in methanol and water.

Again, there is a good linear correlation between the pK_a-values in water in methanol, illustrated in Fig. 5.3 and eq. (5.5).

$$pK_a(MeOH) = 1.08 pK_a(H_2O) + 3.66 \qquad (5.5)$$

There is an obvious trend in Table 5.4 toward larger ΔpK_a-values as the phenols become weaker, reflected numerically in the slope of 1.08 in eq. (5.5). This shows that the phenol pK_a-values, in contrast to those of the carboxylic acids, are slightly more sensitive to substituent effects in methanol than in water. This is not unexpected because the phenoxide oxygen anion is directly conjugated to the aromatic ring, and hence its charge density will vary more

Table 5.4 pK_a-values of phenols in methanol at 25°C[a]

Phenol	pK_a MeOH	pK_aH_2O	ΔpK_a[b]	Phenol	pK_a MeOH	pK_aH_2O	ΔpK_a[b]
2,4,6 trinitro	3.55	0.43	3.12	2-fluoro	12.94	8.73	4.21
2,6-dinitro	7.64	3.74	3.90	3-chloro	13.10	9.02	4.08
2,4-dinitro	7.83	4.10	3.73	4-chloro	13.59	9.38	4.21
2,5-dinitro	8.94	5.22	3.72	4-bromo	13.63	9.36	4.27
2,4,6-tribromo	10.10	6.10	4.00	H	14.33	9.99	4.34
3,5-dinitro	10.29	6.66	3.63	3,5-dimethyl	14.57	10.20	4.37
2-nitro	11.53	7.23	4.30	2-methyl	14.86	10.31	4.55
3,5-dichloro	12.11	8.18	3.93	2,5-dimethyl	14.91	10.41	4.50
3-nitro	12.41	8.36	4.05	2,4-dimethyl	15.04	10.60	4.44
2-chloro	12.97	8.51	4.46	2,4-di-t-butyl	16.77	11.57	5.20

[a] Ref [8]; [b] $\Delta pK_a = pK_a(MeOH) - pK_a(H_2O)$

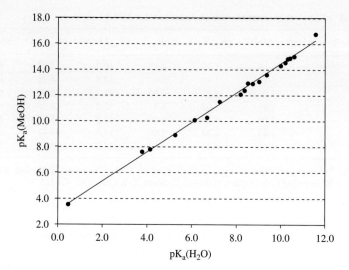

Fig. 5.3.
pK$_a$-values for phenols in methanol versus water at 25°C

strongly with substituent than does that of the carboxylate oxygen atoms. The result is that phenols with more strongly electron-withdrawing substituents are less influenced by solvent change because of the lower charge density on the conjugate phenoxide anion, and hence their pK$_a$-values increase less on transfer to methanol. This effect will be seen to be even more important in the aprotic solvents, where loss of anion solvation plays a much more significant role.

It is noticeable also that the absolute increase in pK$_a$ of phenols on transfer from water to methanol is almost universally less than that of the carboxylic acids, which is again presumably because of the delocalization of the negative charge on the phenoxide anion into the aromatic ring, which results in an overall lower sensitivity to solvent change.

5.2.2 Cationic acids: protonated anilines, amines, N-heterocycles

The dissociation constants for various (protonated) anilines, amines and N-heterocycles in methanol and water are listed in Tables 5.5 and 5.6.

Typically, there is a small increase in pK$_a$ of around 1 unit or less on transfer of the various protonated nitrogen bases from water to methanol, and, as in the case of the neutral acids, there is a good correlation between the values in water in methanol, Fig. 5.4.

The pK$_a$-values for a wide range of nitrogen bases may be represented by eq. (5.6), although the data for anilines is more accurately represented separately by a slightly different line of slope 1.21 and intercept 0.36.

$$pK_a(\text{MeOH}) = 1.02 pK_a(\text{H}_2\text{O}) + 0.72 \qquad (5.6)$$

The modest differences between aqueous and methanolic values for the nitrogen bases are the result of balancing factors: the conversion of R$_3$NH$^+$ to H$^+$ is unfavourable with respect to solvation in methanol relative to water, but the

Table 5.5 pK_a-values of anilinium and ammonium ions in methanol at 25°C[a,b]

Aniline	pK_aMeOH	pK_aH$_2$O	ΔpK_a[c]	Amine	pK_a MeOH	pK_aH$_2$O	ΔpK_a[c]
2-nitro	0.2	−0.25	0.45	hydroxylamine	6.29	5.96	0.33
4-nitro	1.55	0.99	0.55	ammonia	10.78	9.24	1.54
2-bromo	3.46	2.53	0.93	methylamine	11.00	10.64	0.36
2-chloro	3.71	2.67	1.04	ethylamine	11.00	10.60	0.40
3-chloro	4.52	3.51	1.01	butylamine	11.48	10.61	0.87
3-bromo	4.42	3.56	0.86	dimethylamine	11.20	10.72	0.48
4-bromo	4.84	3.88	0.96	trimethylamine	9.80	9.74	0.06
4-chloro	4.95	3.98	0.97	triethylamine	10.78	10.67	0.11
3-methoxy	6.04	4.20	1.84	piperidine	11.07	11.15	−0.08
H	6.05	4.60	1.45	Me$_4$-guanidine	13.20	13.60	−0.40
4-methyl	6.57	5.08	1.49				
4-hydroxy	7.41	5.65	1.76				

[a] Ref [8]; [b] Bos, M., van der Linden, W. E. *Anal. Chim. Acta*, 1996, *332*, 201; [c] ΔpK_a = pK_a(MeOH) − pK_a(H$_2$O)

Table 5.6 pK_a-values of pyridinium ions in methanol at 25°C[a,b]

Pyridine	pK_a MeOH	pK_aH$_2$O	ΔpK_a[c]	Pyridine	pK_a MeOH	pK_a H$_2$O	ΔpK_a[c]
2-chloro	1.00	0.50	0.50	3-acetyl	3.73	3.26	0.47
3-cyano	1.70	1.40	0.30	H	5.44	5.22	0.22
4-cyano	2.03	1.90	0.13	2-methyl	6.18	5.94	0.24
3-chloro	2.83	2.75	0.08	2,6-dimethyl	6.86	6.68	0.18
3-bromo	2.90	2.80	0.10	4–N, N–Me$_2$	10.10	9.60	0.50

[a] Ref [8]; [b] Augustin-Nowacka, D., Makowski, M., Chmurzynski, L. *Anal. Chim. Acta*, 2000, *418*, 233; [c] ΔpK_a = pK_a(MeOH) − pK_a(H$_2$O)

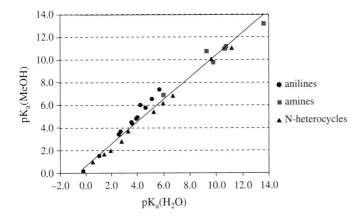

Fig. 5.4.
pK_a-values for protonated nitrogen bases in methanol versus water at 25°C

increased stability of the neutral free-base in methanol increases the tendency towards dissociation of R_3NH^+ in methanol.

5.2.3 Summary

- pK_a-values for carboxylic acids, phenols and protonated amines are higher in methanol than in water, but the increases are much larger for the neutral acids
- The behaviour of carboxylic acids, phenols, and amines in methanol closely parallels that in water. Good linear correlations between aqueous and methanolic values, of the form shown in eq. (5.7), are observed.

$$pK_a(MeOH) = mpK_a(H_2O) + c \tag{5.7}$$

The best-fit values of m and c for the different acids are:

Acid type	m	c
carboxylic acid	1.02	4.98[a]
phenol	1.08	3.66
protonated nitrogen base[b]	1.02	0.72

[a] For the second pK_a of dicarboxylic acids, $c = 6.2$; [b] A correlation based on anilines alone gives $m = 1.21$ and $c = 0.38$

- Accurate prediction of pK_a-values in methanol from those in water follow from eq. (5.7)

5.3 Higher alcohols

The solvation of both anions and cations ions decreases in the order MeOH > EtOH > i-PrOH >> t-BuOH (Section 3.3.1), and this is expected to be reflected in increased pK_a-values of carboxylic acids and phenols, in particular, across this series. Details of conductimetric (ion-pair formation) and potentiometric measurements in i-propanol and t-butanol have been described by Kolthoff and Chantooni [11–13].

Table 5.7 lists representative dissociation constants for carboxylic acids and phenols in the higher alcohols [11–18].

Fig. 5.5 shows a plot of the pK_a-values for carboxylic acids in the alcohols against those in water (values in n-BuOH have been omitted for clarity, but are similar to those in EtOH).

It is apparent from Table 5.7 and Fig. 5.5 that the increases in pK_a from methanol to ethanol, i-propanol and n-butanol are relatively modest, but that significant increases occur on transfer to t-butanol. The correlation involving results in t-butanol is somewhat more scattered, but there is certainly an increase in both the absolute values and the slope compared with the other alcohols. Furthermore, homohydrogen-bond formation is no longer negligible

Table 5.7 pK_a-values of carboxylic acids and phenols in alcohols at 25°C[a]

Carboxylic acid/Phenol	pK_a MeOH	pK_a EtOH	pK_a i-PrOH	pK_a n-BuOH	pK_a t-BuOH
dichloroacetic	6.38	6.89	7.8	7.45	10.27
cyanoacetic	7.50	8.00		8.01	10.68
chloroacetic	7.88	8.45	9.23	8.49	12.24
acetic	9.72	10.44	11.35		14.60
2-nitrobenzoic	7.64	8.26			
3,5-dinitrobenzoic	7.38		8.31		10.6
3-nitro,4-chlorobenzoic			9.34		11.75
4-nitrobenzoic	8.34	8.90	9.60		12.04
3,4-dichlorobenzoic	8.53		9.82		12.97
3-bromobenzoic	8.80	9.42	10.11		13.48
4-chlorobenzoic	9.04	9.69			
benzoic	9.30	10.13	11.75	10.24	15.1
2,4,6-trinitrophenol	3.55		4.02		5.35
4-nitrophenol	11.30		12.45		15.88
3-nitrophenol	12.41		13.92		16.99
3-bromophenol			14.83		18.88
4-chlorophenol			15.31		18.96
4-bromophenol	13.63		15.36		18.88

[a] Ref [11–18]

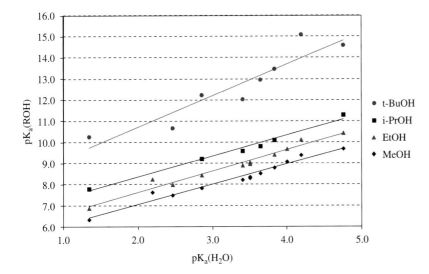

Fig. 5.5.
pK_a-values of carboxylic acids in alcohols versus water at 25°C

in t-BuOH; for example $K_{AHA} = 30M^{-1}$ for 3,5-dinitrophenol [13]. Both effects are compatible with reduced anion solvation being a more dominant feature in t-butanol, as this will most strongly affect the weakest acids which have the highest charge densities on the anions. The scatter in the t-butanol results might at least partially reflect the greater difficulties associated with measurements in this solvent, because of extensive ion-pair (K_{IP} typically $> 10^4 M^{-1}$) and homohydrogen-bond formation [9, 11–13, 18].

Table 5.8 Free energies of transfer of aromatic carboxylic acids between methanol and higher alcohols at 25°C[a]

$\Delta G_{tr}(HA)/kJmol^{-1}$

Acid	MeOH[b]	EtOH	i-PrOH	t-BuOH
benzoic	0			1.34
3-bromobenzoic	0		1.31	0.81
4-bromobenzoic	0		0.39	−0.23
3,4-dichlorobenzoic	0		−0.91	−1.72
4-nitrobenzoic acid	0	2.15	3.41	1.97

[a] Ref [13]; [b] Solubility of acids in MeOH: 3.16M, 1.51M, 0.12M, 0.27M, 0.20M, respectively

Table 5.9 pK_a-values of HCl and HBr in alcohols at 25°C[a]

Acid	$pK_a(H_2O)$	$pK_a(MeOH)$	$pK_a(EtOH)$	$pKa(i\text{-PrOH})$	$pK_a(t\text{-BuOH})$
HCl	(−5.5)	1.1	2.0	3.1	5.5
HBr	(−6.5)	0.8	1.7	2.0	5.0

[a] Ref [19]

There is little change in the free energies of the neutral substrates among the alcohols; some representative values for carboxylic acids are given in Table 5.8.

The absolute solubility of the individual acids in any given solvent varies by more than an order of magnitude, but the change among the solvents is relatively small; it corresponds on average to $\Delta G_{tr}(HA) = 1.4 \text{kJ mol}^{-1}$ on transfer from MeOH, equivalent to ~ 0.25 pK units, with a maximum effect of ~ 0.5 units.

The dissociation constants of HCl and HBr have also been reported in several alcohols, and the results are listed in Table 5.9 [19].

The acids are still relatively strong in MeOH, but become increasingly weaker in the higher alcohols. The dominant effect is the decreased solvation of the halide ions in the higher alcohols relative to water. In t-BuOH, significant formation of HX_2^- occurs, and it is possible to define an alternative acidity, given by eq. (5.8).

$$2HX \xrightleftharpoons{K(2HX)} H^+ + HX_2^- \quad (5.8)$$

pK(2HX) values for HCl and HBr, respectively, in t-BuOH are 4.7 and 3.7.

5.4 Alcohol–water mixtures*

*See Appendix 3.1 for a discussion of composition scales in mixed solvents

Measurement of dissociation constants in alcohol–water mixtures using electrochemical measurements based on the glass electrode is relatively straightforward, and a particularly convenient method due to Grunwald and co-workers [20] has been described earlier (Section 4.5.1).

The dissociation of carboxylic acids, such as acetic acid and benzoic acid, shows strong evidence of selective solvation of the carboxylate and hydrogen ions by water in the mixtures (Section 3.3.3). This is illustrated by the results in Fig. 5.6 for acetic acid in methanol–water and ethanol–water mixtures [20–22], which show that the pK_a-values in 60 wt% ROH have increased by only around 1.2 units relative to those in water compared with a total of 5–5.5 on transfer to the pure alcohol.

Similar results are observed in other organic–water mixtures (Chapter 8), such as acetonitrile–water [23, 24], dioxane–water [25], THF–water [26], and DMF–water [27]. For example, acetic acid in acetonitrile has $pK_a = 22.9$ (i.e., some 18.3 units higher than in water), but its pK_a increases by less than two units to 6.75 on transfer to 60 wt% acetonitrile [23]. Indeed, for acetic acid, pK_a-values in aqueous mixtures containing up to 60 wt% organic component are almost independent of the nature of the organic component and the value in the pure solvent. Similar trends are also observed for a series of phenols in the mixtures. These results all show that solvation of the ions by water remains dominant in the mixtures.

The dissociation of ammonium and anilinium ions are also subject to preferential solvation, but the consequences are somewhat different to those observed for the carboxylic acids and phenols [5, 20]. Typical behaviour for these cationic acids is illustrated in Fig. 5.7 for the anilinium ion in ethanol–water mixtures [5].

The addition of ethanol to water results in an initial decrease in the pK_a of the anilinium ion, which persists until about 80 wt% ethanol; beyond this there is a rapid increase to the value in pure ethanol, $pK_a = 5.7$, as the remaining water is removed. The net effect is that the anilinium ion is a stronger acid across most of the range of mixtures than in either of the pure component solvents.

The reason for this is that the initial addition of ethanol has little influence on the ions, which are preferentially solvated by water, but stabilizes the free base, aniline, thereby increasing the tendency of the anilinium ion to dissociate. This effect continues with increasing addition of ethanol until most of the water has been replaced, at which point the decreased solvation of the proton

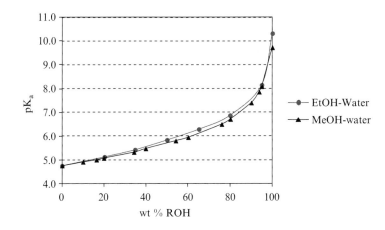

Fig. 5.6.
pK_a-values for acetic acid in MeOH-water and EtOH-water at 25°C [20–22]

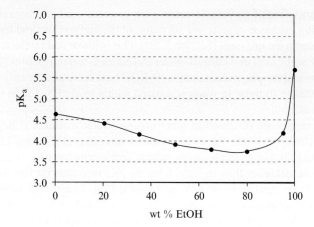

Fig. 5.7.
pK$_a$-values for the anilinium ion in EtOH-water at 25°C [5]

predominates and causes a net increase in pK$_a$. Other amines show similar behaviour. The same general trends occur in acetonitrile–water mixtures [28]; the anilinium and ammonium ions are much weaker acids in pure acetonitrile than in water, but addition of acetonitrile to water causes an increase in acidity.

A comprehensive discussion of the acidity of carboxylic acids, phenols, amines and pyridine derivatives in methanol–water mixtures is given by Bosch, Rosés, and co-workers [29].

5.5 Salt formation in alcohols and aqueous–alcohol mixtures

The protonation of amines by carboxylic acids to form the corresponding ammonium salts becomes progressively more difficult as the alcohol content of the alcohol–water mixtures increases. The equilibrium constant, log K$_e$, for protonation of aniline by acetic acid, eq. (5.9), for example, in ethanol–water mixtures shows a monotonic and almost linear decrease with ethanol content, with an overall change of 4.5 log units, Fig. 5.8 [5, 22], despite the more complex influence of preferential solvation on the individual curves for the acids and bases noted above (Fig. 5.6 and 5.7).

$$CH_3CO_2H + ArNH_2 \xrightleftharpoons{K_e} CH_3CO_2^- + ArNH_3^+ \qquad (5.9)$$

The explanation for this steady decrease in log K$_e$ with added ethanol is apparent from an analysis of the influence of solvent change on the species in eq. (5.9). The initial decrease on K$_e$ is a result of increased of solvation of aniline, which makes it more difficult to protonate. The effect on aniline levels off at higher ethanol content, but at these lower water levels the large reduction in the solvation of the acetate ion in particular further inhibits the proton transfer reaction. The combined result of more efficient solvation of aniline and poorer solvation of the ions as the ethanol content of the mixtures increase is an overall decrease in K$_e$ of more than four orders of magnitude.

Salt formation in alcohols and aqueous–alcohol mixtures

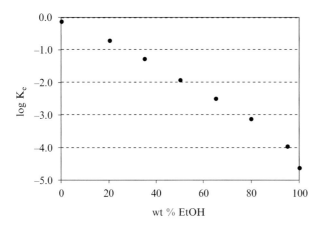

Fig. 5.8.
Equilibrium constant for protonation of aniline by acetic acid in EtOH-water mixtures at 25°C [5, 22]

The results in Fig. 5.8 are calculated from dissociation constants determined under ideal (dilute) conditions. In solutions containing higher concentrations of acetic acid and aniline, however, it is necessary to include ion-pair formation between the carboxylate and ammonium ions as part of the overall equilibrium; this can influence the equilibrium position quite strongly in the direction of increased ionization as the solvent composition approaches that of pure ethanol (Chapter 8).

A consequence of the change in equilibrium position between carboxylic acid and amine groups may also be seen in zwitter-ion formation of amino acids, such as 4-aminobenzoic acid, which involves *intramolecular* proton transfer.

In keeping with the results illustrated in Fig. 5.8, the equilibrium proportion of the zwitter-ion decreases strongly as the ethanol content of ethanol–water mixtures increases, Table 5.10 [30].

The extent of zwitter-ion formation is thus almost 1,000 times lower in 75 wt% ethanol–water than in pure water.

Table 5.10 Zwitter-ion formation of 4-aminobenzoic acid in ethanol-water mixtures at 25°C[a]

	Water	50%EtOH-H_2O	75%EtOH-H_2O
% zwitter ion	10.5	0.27	0.015
log K_{zi}	−0.94	−2.57	−3.83

[a] Ref [30]

5.6 Formamide, acetamide, N-methylpropionamide

Formamide is a highly polar solvent (Section 1.2) with a dielectric constant larger than that of water ($\varepsilon_r = 109.5$ at 25°C), a large dipole moment, and relatively high values of both Donor and Acceptor Numbers. It has an autoionization constant of $pK_{ai} = 16.8$ [1], similar to that of methanol (Table 5.1).

formamide   acetamide   N-methylpropionamide

These properties result from a combination of a polar carbonyl group and relatively acidic N–H protons, which can readily take part in H-bond formation with anions. They are reflected in the high solubility of simple salts, such as NaCl and KCl (1.5 M and 0.8 M, respectively) [31], which are approaching those in water. Similar properties are also shared by acetamide and N-methylpropionamide, although we note that the high melting point of acetamide (81°C) means that it can only be used as a solvent at elevated temperatures.

The favourable combination of high polarity and H-bond acidity exhibited by these amides suggests that dissociation constants for both neutral and cationic acids, when compared to other non-aqueous solvents, might be expected to be closer to those in water.

Nayak, Dash, and co-workers have reported dissociation constants for a range of carboxylic acids and anilinium ions in formamide [32–37]. Measurements were performed electrochemically using cells of the type described in Chapter 4, eq. (4.4), or its equivalent, in which a quinhydrin electrode is used in place of the hydrogen electrode to determine the hydrogen ion activity. Homohydrogen-bond formation and ion association are not important, and activity coefficients do not differ significantly from unity at concentrations below 0.1M (Table 4.1).

Dissociation constants for carboxylic acids in formamide are listed in Table 5.11.

The pK_a-values show a small average increase of ~ 2.3 pK units relative to water, equivalent to an increase in the free energy of dissociation of 13 kJ mol^{-1}; this may be compared with the corresponding values for methanol of $\Delta pK_a = 5.0$ (28.5 kJ mol^{-1}). The decrease in anion solvation on transfer from water to the two solvents is similar, e.g., $\Delta G_{tr}(OAc^-) = 20$ kJ mol^{-1} in formamide compared with 16 kJ mol^{-1} in methanol (Tables 3.5 and 3.6), but the proton is more stable in formamide than in methanol. The result is that dissociation constants for carboxylic acids in formamide (and acetamide, below) are the closest of any non-aqueous solvents to those in water.

Anilinium ions, Table 5.12, exhibit pK_a-values which are on average around 0.5 units higher than those in water.

Table 5.11 pK_a-values of carboxylic acids in formamide ($HCONH_2$) at 25°C[a]

Aliphatic acid	$pK_a HCONH_2$	$pK_a H_2O$	ΔpK_a[b]	Benzoic acid	$pK_a HCONH_2$	$pK_a H_2O$	ΔpK_a[b]
2-chloroacetic	4.10	2.87	1.23	3,5-dichloro	6.08	3.11	2.97
lactic	5.81	3.86	1.95	4-nitro	5.88	3.44	2.44
phenylacetic	6.57	4.31	2.26	3-nitro	5.40	3.51	1.89
acetic	6.91	4.76	2.15	3-chloro	6.22	3.70	2.52
propanoic	7.26	4.87	2.39	3-bromo	6.22	3.70	2.52
butanoic	7.34	4.82	2.52	3-iodo	6.27	3.85	2.42
				4-iodo	6.47	3.93	2.54
Benzoic acid:				4-chloro	6.58	3.99	2.59
2-nitro	4.41	2.17	2.24	4-bromo	6.53	4.00	2.53
3,5-dinitro	4.45	2.75	2.24	H			
2-bromo	5.86	2.85	3.01	3-methyl	6.42	4.28	2.14
2-chloro	5.82	2.92	2.90	4-methyl	6.82	4.37	2.45

[a] Data from ref [32–37]; [b] $\Delta pK_a = pK_a(HCONH_2) - pK_a(H_2O)$

Table 5.12 pK_a-values of anilinium ions in formamide ($HCONH_2$) at 25°C[a]

Aniline	$pK_a HCONH_2$	$pK_a H_2O$	ΔpK_a[b]	Aniline	$pK_a HCONH_2$	$pK_a H_2O$	ΔpK_a[b]
4-nitro	1.77	1.02	0.75	4-bromo	4.29	3.89	0.40
2-bromo	3.00	2.53	0.47	H	5.10	4.58	0.52
3-nitro	3.05	2.50	0.55	N-methyl	5.26	4.85	0.41
3-bromo	3.96	3.53	0.43				

[a] Data from ref [35–37]; [b] $\Delta pK_a = pK_a(HCONH_2) - pK_a(H_2O)$

Dissociation constants of protonated aliphatic amines in formamide have not been reported, but as a first approximation pK_a-values could be estimated by adding 0.5 units to the corresponding aqueous values.

Limited data for neutral acids in acetamide [38], Table 5.13, suggests that pK_a-values are ~ 1 unit higher than the corresponding values in water.

Finally, a very precise measurement of the pK_a of acetic acid in N-methylpropionamide [39], determined by EMF measurements on cells with hydrogen gas and silver–silver chloride electrodes without liquid junction, give a value of $pK_a = 7.995$, compared with 4.756 in water. The difference of 3.24 is larger than the corresponding value in formamide ($\Delta pK_a = 2.15$), despite

Table 5.13 pK_a-values of neutral acids in acetamide (CH_3CONH_2) at 98°C[a]

Acid	$pK_a CH_3CONH_2$	$pK_a H_2O$	ΔpK_a[b]	Acid	$pK_a CH_3CONH_2$	$pK_a H_2O$	ΔpK_a[b]
dichloroacetic	2.0	1.3	0.7	4-chlorophenol	9.85		
formic	4.8	3.95	0.85	phenol	10.35	9.1	1.3
acetic	5.8	4.76	1.0	phosphoric(pK_1)	3.8	2.5	1.3
benzoic	5.55	4.38	1.17	phosphoric(pK_2)	8.1	7.4	0.7

[a] Data from ref [38]; [b] $\Delta pK_a = pK_a(CH_3CONH_2) - pK_a(H_2O)$

the very high dielectric constant of NMP ($\varepsilon_r = 176$). This further emphasises the importance of specific solvation effects in determining pK_a-values.

In summary, the polar, protic primary and secondary amide solvents stabilize ions more effectively than methanol, especially the proton, but pK_a-values of neutral and cationic acids are still somewhat higher than those in water. The trends observed in Tables 5.11–5.13 can be used as a basis for estimating dissociation constants for other substrates in these solvents.

5.7 Formic acid

Formic acid, although not widely used as a solvent, is of interest as an example of a protic solvent with high acidity. Among its solvent properties are a relatively high dielectric constant, $\varepsilon_r = 56.1$, and dipole moment, $\mu = 1.4$ Debye, but a weak hydrogen-bond basicity (Section 1.2), $\beta = 0.38$ [40, 41]. It also has a very high autoionization constant, $pK_{ai} = 6.2$ [42], which means that all but the most weakly basic amines will be quantitatively protonated in formic acid.

Dissociation constants of several acids in formic acid (Table 5.14) have been measured by a combination of electrochemical and spectrophotometric methods [42, 43], with good agreement between results obtained by the two methods. Glass electrode potentials in formic acid were found to be stable and reproducible. Spectrophotometric measurements were based upon influence of the acids on the ionization of various acid–base indicators: o-dinitrodiphenylamine ($pK_a = 1.3$), bromocresol green ($pK_a = 3.9$) and bromothymol blue ($pK_a = 5.2$).

Most of the acids in Table 5.14 have pK_a-values which are similar to or slightly higher than the corresponding values in methanol (Section 5.2), and several units higher than those in water. This suggests that, compared with methanol, the effect of the expected more favourable solvation of anions in formic acid is more than compensated by the weaker interaction of the proton with formic acid, because of its low H-bond basicity.

The most striking result is, however, that for benzoic acid, which has a pK_a-value in formic acid that is 4.5 units lower than in methanol ($pK_a = 9.38$) and approaching that in water ($pK_a = 4.19$). This must reflect an especially strong stabilization by hydrogen bonding of the benzoate anion by formic acid; presumably similar results would obtain for other carboxylic acids.

Table 5.14 pK_a-values of acids in formic acid at 25°C[a]

Acid	pK_a	Acid	pK_a
HBr	0.4	CF_3CO_2H	4.65
HCl	1.45	Picric acid	4.85
CH_3SO_3H	1.60	Benzoic acid	4.90
p-TSA[b]	1.65	HF	4.96[c]
HSO_4^-	4.30		

[a] Ref [42]; [b] p-toluenesulfonic acid; [c] Ref [43]

References

[1] Rondinini, S., Longhi, P., Mussini, P. R., Mussini, T. *Pure & Appl. Chem.*, 1987, *59*, 1693
[2] Bates, R. G., 'Determination of pH – Theory and Practice', 2nd edn. Wiley, New York, 1973
[3] Parsons, G. H., Rochester, C. H. *J. Chem. Soc., Faraday Trans* 1, 1972, *68*, 523
[4] Koskikallio, J. *Suomen Kem.*, 1957, *30B*, 111
[5] Gutbezahl, B., Grunwald, E. *J. Amer. Chem. Soc.*, 1953, *75*, 559
[6] Rochester, C. H. *J. Chem. Soc., Dalton Trans.*, 1972, 5
[7] Reference 2, Chapter 8
[8] Rived, F., Rosés, M., Bosch, E. *Anal. Chim. Acta*, 1998, *374*, 309
[9] Chantooni, M. K., Kolthoff, I. M. *J. Phys. Chem.*, 1974, *78*, 839
[10] Chantooni, M. K., Kolthoff, I. M. *J. Phys. Chem.*, 1975, *79*, 1176
[11] Chantooni, M. K., Kolthoff, I. M. *J. Phys. Chem.*, 1978, *82*, 994
[12] Kolthoff, I. M., Chantooni, M. K. *J. Am. Chem. Soc.*, 1965, *87*, 4428
[13] Chantooni, M. K., Kolthoff, I. M. *Anal. Chem.*, 1979, *51*, 133
[14] Minnick, J. L., Kilpatric, M. *J. Phys. Chem.*, 1939, *43*, 259
[15] Dash, U. N., Kalia, S. P., Mishra, M. K. *J. Electrochem. Soc. India*, 1988, *7*, 323
[16] Mason, B. R., Kilpatric, M. *J. Amer. Chem. Soc.*, 1937, *59*, 572
[17] Bosch, E, Ràfols, C., Rosés, M. *Talanta*, 1989, *36*, 1227
[18] Bosch, E., Rosés, M. *Talanta*, 1989, *36*, 627
[19] Kolthoff, I. M., Chantooni, M. K. *J. Phys. Chem.*, 1979, *83*, 468
[20] Bacarella, A. L., Grunwald, E., Marshall, H. P., Purlee, E. L. *J. Org. Chem.*, 1955, *20*, 747
[21] Bacarella, A. L., Grunwald, E., Marshall, H. P., Purlee, E. L. *J. Phys. Chem.*, 1958, *62*, 856
[22] Grunwald, E., Berkowitz, B. J. *J. Am. Chem. Soc.*, 1951, *73*, 4939
[23] Espinosa, S., Bosch, E., Rosés, M. *J. Chromatog. A*, 2002, *964*, 55
[24] Barbosa, J., Beltrán, J. L., Sanz-Nebot, V. *Anal. Chim. Acta.*, 1994, *288*, 271
[25] Harned, H. S., Kazanjian, G. L. *J. Am. Chem. Soc.*, 1936, *58*, 1912
[26] Barbosa, J., Barrón, D., Butí, S. *Anal. Chim. Acta*, 1999, *389*, 31
[27] González, A. G., Rosales, D., Gómez-Ariza, J. L., Sanz, J. F. *Anal. Chim. Acta*, 1990, *228*, 301
[28] Espinosa, S., Bosch, E., Rosés, M. *Anal. Chim. Acta*, 2002, *454*, 157
[29] Rived, F., Canals, I., Bosch, E., Rosés, M. *Anal. Chim. Acta*, 2001, *439*, 315
[30] Van de Graaf, B., Hoefnagel, A. J., Wepster, B. M. *J. Org. Chem.*, 1981, *46*, 653
[31] Vincent, C. A. Solubility Data Series, 1980, *11*, 9–13; 28–34
[32] Dash, U. N., Nayak, S. K., Nayak, B. *J. Electrochem. Soc. India*, 1995, *44*, 37
[33] Dash, U. N. *Thermochim. Acta*, 1979, *32*, 33
[34] Dash, U. N. *Thermochim. Acta*, 1978, *27*, 379
[35] Acharya, K., Nayak, B. *Electrochim. Acta*, 1983, *28*, 1041
[36] Acharya, K., Nayak, B. *Thermochim. Acta*, 1983, *66*, 377
[37] Dash, U. N., Nayak, B. *Aust. J. Chem.*, 1975, *28*, 797
[38] Guiot, S., Tremillon, B. *J. Electroanal. Chem.*, 1968, *18*, 261
[39] Etz, E. S., Robinson, R. A., Bates, R. G. *J. Soln. Chem.*, 1972, *1*, 507
[40] Oudsema, J. W., Poole, C. F. *J. High Resolution Chromatogr.*, 1993, *16*, 130
[41] Chialvo, A. A., Kettler, M., Nezbeda, I. *J. Phys. Chem., B*, 2005, *109*, 9736
[42] Breant, M., Beguin, C., Coulombeau, C. *Anal. Chim. Acta*, 1976, *87*, 201
[43] Coulombeau, C., Beguin, C., Coulombeau, C. *J. Fluorine Chem.*, 1977, *9*, 483

6 High-Basicity Polar Aprotic Solvents

Polar aprotic solvents are used in around 10% of chemical manufacturing processes and in numerous laboratory procedures, ranging from synthetic to mechanistic studies. They are universally poor at solvating anions, i.e., at accepting electrons, but show wide variations in their ability to interact with cations. Hence the most useful means of classifying them is through their differing ability to solvate cations, and more particularly the proton, via electron donation. We have chosen as a convenient measure of this the solvent *Donor Numbers*, DN (Section 1.2.1) [1], based on the enthalpy of adduct formation between the solvent molecule and $SbCl_5$, but other commonly used measures of Lewis basicity or hydrogen-bond-acceptor basicity [2–4] would serve equally well for this purpose.

On this basis, the aprotic solvents fall into two convenient groups, the members of each of which have closely related, and hence predictable, effects on dissociation constants. The first of these comprise solvents with high Donor Numbers, and includes dimethylsulphoxide (DMSO), *N*,*N*-dimethylformamide (DMF), *N*-methylpyrrolidin-2-one (NMP), *N*,*N*-dimethylacetamide (DMAC), and liquid ammonia; the most comprehensively studied of these is DMSO [5–7]. Hexamethylphosphoramide (HMPT), which has the highest Donor Number and Lewis basicity amongst all of the common solvents considered, also belongs to this group but is now rarely used because of concerns over its potential carcinogenicity [8].

The second group, characterized by low Donor Numbers, includes acetonitrile (MeCN), 4-methyl-1,3-dioxolane (propylene carbonate, PC), nitromethane (NM), acetone, tetrahydrofuran (THF), tetrahydrothiophene 1,1-dioxide (sulpholane, TMS), and methyl isobutyl ketone (MIBK). Amongst this group, the largest body of data reported relates to measurements in MeCN [9–12]. The properties of acids and bases in this group of solvents are discussed in Chapter 7.

All of the aprotic solvents have very low autoionization constants (Section 4.6), with typically $pK_{ai} > 30$, although exact values of K_{ai} are often uncertain due to experimental difficulty in their measurement. This arises in part because of the strong sensitivity of the results obtained by electrochemical measurements in these media to trace impurities. The most reliable value is probably that for the ionization of DMSO, $pK_{ai} = 35.1$, determined by Bordwell and co-workers [13]. For MeCN, a lower limit of $pK_{ai} = 33.3$ has been reported [14]. The very high pK_{ai}-values mean that the solvents can tolerate very strong bases.

A further important consequence of the poor ability of aprotic solvents to solvate anions is that the dissociation of neutral acids is normally strongly influenced by homohydrogen-bond formation, eq. (6.1) (Section 4.4).

$$RCO_2H \underset{}{\overset{K_{HA}}{\rightleftharpoons}} H^+ + RCO_2^-$$

$$RCO_2H + RCO_2^- \underset{}{\overset{K_{AHA}}{\rightleftharpoons}} RCO_2^- \cdots HO_2CR \qquad (6.1)$$

K_{AHA} for carboxylic acids in DMSO, for example, are typically around 50 M^{-1} (Chapter 5, Table 4.2), and hence hydrogen-bond association between the acid and anion is important at concentrations above 10^{-3} M; simultaneous solution of the equations representing the two equilibria is thus often necessary in order to calculate the species distribution in a solution of given pH. Failure to recognize or allow for these complications and also, in the case of low polarity solvents, ion association, can lead to significant errors in pK$_a$-determinations. Where discrepancies between dissociation constants determined in different laboratories exist, we have chosen those for which adequate allowance has been made for homohydrogen-bond formation and also ion-association, particularly those from Bordwell, Kolthoff, and co-workers.

In practical applications, such as product isolation through salt formation, or the ionization of carbon acids in important synthetic procedures, we are often primarily interested in acid–base equilibria, rather than the ionization of individual acids, e.g.:

$$HA + B \rightleftharpoons BH^+ + A^-$$

$$HA_1 + A_2^- \rightleftharpoons HA_2 + A_1^-$$

Importantly, the solvated proton, which is mainly responsible for the differences between the acidities in various aprotic solvents, is no longer involved. In principle, the equilibrium constant for such reactions follows directly from the difference in pK$_a$-values of the two acids, but hydrogen-bond association and ion-pair formation between the components may have a profound influence on the overall equilibria, especially at concentrations relevant to synthetic procedures. Detailed discussion of such equilibria is presented in Chapter 8.

A comprehensive listing of dissociation constants in polar aprotic solvents measured prior to 1990 is given by Izutsu [15], and in this and the following chapter we update the results and consider relationships exhibited by the data within individual solvents and among the various solvents.

6.1 Dimethylsulphoxide

Bordwell and co-workers have reported dissociation constants for more than 300 substrates in DMSO using a spectrophotometric method based on a series of overlapping indicators, described in Section 4.5.2 [5, 16]. The potentiometric method, employing the glass electrode (Section 4.5.1), has also been applied to the determination of acidities, including those of carboxylic acids and phenols [6, 7]. More extensive tabulations of data in dimethylsulphoxide are included in Appendix 9.2.

6.1.1 Neutral acids: carboxylic acids, phenols and thiophenols, water and methanol, anilines and amides, carbon acids

Carboxylic acids

Dissociation constants for a variety of representative aliphatic and aromatic carboxylic acids [7, 17–20] are listed in Table 6.1 together with the corresponding values in water.

For the acids listed in Table 6.1, there is an average increase in pK_a on transfer from water to DMSO of $\Delta pK_a = 6.2 (\equiv 35.3\,\text{kJ mol}^{-1})$, but this broad number conceals a systematic trend towards larger ΔpK_a as the acid becomes weaker. A plot of $pK_a(\text{DMSO})$ against $pK_a(\text{H}_2\text{O})$ is shown in Fig. 6.1.

There is some scatter in the correlation, with individual deviations of up to 0.5 pK units being observed; these are often greater than the experimental

Table 6.1 pK_a-values of carboxylic acids in dimethylsulfoxide at 25°C[a]

Aliphatic acid	pK_a DMSO	$pK_a\text{H}_2\text{O}$	ΔpK_a[b]	Benzoic acid	pK_a DMSO	$pK_a\text{H}_2\text{O}$	ΔpK_a[b]
dichloroacetic	6.4	1.34	5.1	3-nitro	9.2	3.47	5.7
2-chloroacetic	8.9	2.85	6.0	3,5-dichloro	8.8	3.56	5.3
acetic	12.6	4.76	7.9	3-bromo	9.7	3.83	5.9
butanoic	12.9	4.86	8.0	4-bromo	10.5	3.99	6.5
				4-chloro	10.1	4.00	6.1
Benzoic acid:				H	11.1	4.19	6.9
2,4-dinitro	6.5	1.43	5.1	3-methyl	11.0	4.28	6.7
2-nitro	8.2	2.19	6.0	4-methyl	11.2	4.38	6.8
2-chloro	9.3	2.92	6.4	3,4-dimethyl	11.4	4.41	7.0
4-nitro	9.0	3.43	6.6	4-hydroxy	11.8	4.55	7.2

[a] Ref [7, 17–20]; [b] $\Delta pK_a = pK_a(\text{DMSO}) - pK_a(\text{H}_2\text{O})$

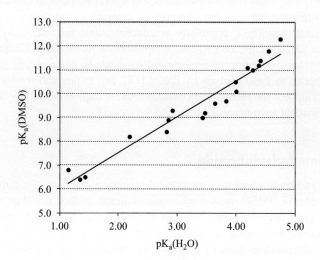

Fig. 6.1.
pK_a-values of carboxylic acids in dimethylsulfoxide versus water at 25°C

uncertainties. The scatter appears to be primarily related to specific solvation effects in water, rather than in DMSO, as it is largely absent in correlations between aromatic pK_a-values among various pairs of non-aqueous solvents, e.g., DMSO versus MeOH or DMF.

The best-fit line shown in Fig. 6.1 corresponds to eq. (6.2).

$$pK_a(DMSO) = 1.57 pK_a(H_2O) + 4.21 \qquad (6.2)$$

The slope of the plot is significantly greater than unity, which means that the dissociation constants are more sensitive to substituents in DMSO than in water. The practical consequence is that whereas the weaker acids increase by more than 7 pK units in DMSO relative to water, the corresponding increase for the stronger acids is less than 5 units.

An analysis of the changes in pK_a in terms of the influence of solvent on the individual components can be achieved by application of eq. (6.3) (Chapter 3, eq. (3.4)) to the data.

$$\Delta pK_a = \{\Delta G_{tr}(H^+) + \Delta G_{tr}(A^-) - \Delta G_{tr}(HA)\}/2.303RT \qquad (6.3)$$

The aromatic acid molecules show an average decrease in free energy of $\Delta G_{tr}(HA) = -21.4\,\text{kJ mol}^{-1}(\Delta pK_a = 3.8)$ on transfer from water [21, 22], but the increased solvation of the proton (Table 3.7), $\Delta G_{tr}(H^+) = -19.6\,\text{kJ mol}^{-1}(\Delta pK_a = -3.4)$ largely counterbalances the effect of this on the dissociation constants. The *dominant contribution* to the observed change in pK_a results from the decreased in solvation of the carboxylate anions in DMSO. $\Delta G_{tr}(A^-)$ values show increases ranging from $25.7\,\text{kJ mol}^{-1}(\Delta pK_a = 4.5)$ for the stronger acids, to $36.8\,\text{kJ mol}^{-1}(\Delta pK_a = 6.5)$ for the weaker acids. This variation from stronger to weaker acids parallels the increase in the charge density of the carboxylate oxygen atoms as the acids become weaker. The strongest acids, for example, correspond to those whose conjugate anions have the lowest charge density on the carboxylate oxygen atoms, and hence the weakest hydrogen-bonding interaction with water; thus they show smaller increases in free energy on transfer to DMSO.

The dissociation of dicarboxylic acids in DMSO, illustrated for malonic acid, eq. (6.4), can be strongly influenced by intramolecular H-bonding within the mono-anion, eq. (6.5) [18].

Intramolecular H-bond formation in water and to a lesser extent methanol is normally very weak, because of strongly competing intermolecular bonding between the solvent and both the carboxylic and carboxylate groups, but in

DMSO intermolecular H-bonding is limited to that of the carboxylic acid group with DMSO:

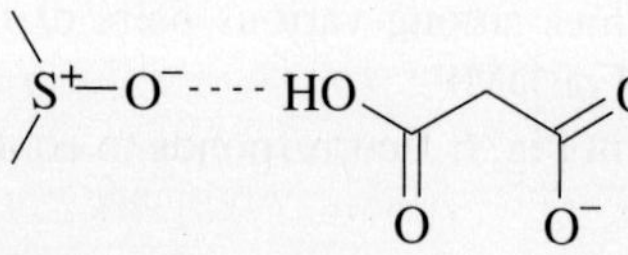

The influence of intramolecular H-bonding (eq. (6.5)) on the acid dissociation equilibria is therefore much stronger than in protic solvents.

Representative acids, with increasing numbers, n, of $-CH_2$ groups separating the carboxylic acids ($n = 0, 1, 2, 3, 4, 7$), are listed in Table 6.2, along with the dissociation constants for the first and second dissociations.

The equilibrium constant for the complete ionization of the dicarboxylic acids is shown in Scheme 6.1, the equilibrium constant for which is given by $K_e = K_{a1}K_{a2}$. Thus we may represent the total increase in pK_a across the two steps by $\Delta pK_a = \Delta pK_{a1} + \Delta pK_{a2}$.

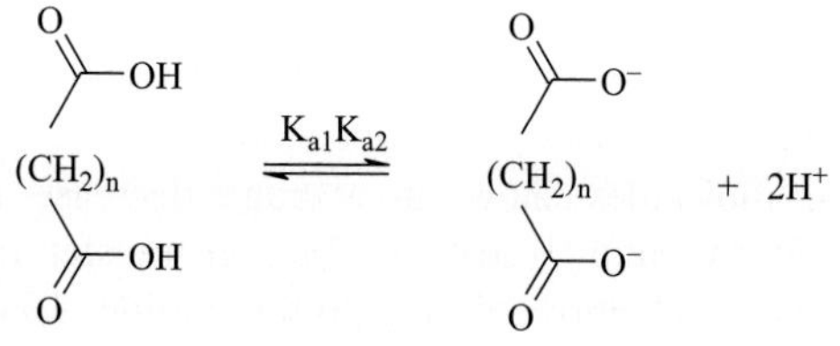

Scheme 6.1.
Complete ionization of dicarboxylic acids

It follows from the data in Table 6.2 that ΔpK_a is relatively constant across the whole series of acids, averaging around 16.3 units, reflecting predominantly the very poor solvation of the dicarboxylate anion in DMSO, which is only partially compensated for by the increased stability of the proton. The division between the two dissociation steps is, however, very substrate-dependent. For the longer-chain acids, adipic ($n = 4$) and azelaic ($n = 7$) acids, the successive dissociation processes are largely independent of one another, and pK_{a1} and pK_{a2} show increases approaching 8 units each. Furthermore, the increases in the first dissociation constant, ΔpK_{a1}, are similar to those of their mono-esters and of simple mono-carboxylic acids, such as acetic and butanioc acids. By contrast, for malonic acid ($n = 1$), strong, intramolecular H-bond stabilization of the mono-anion, eq. (6.5), reduces ΔpK_{a1} sharply and correspondingly

Table 6.2 pK_a-values of aliphatic dicarboxylic acids in dimethylsulfoxide at 25°C[a]

Dicarboxylic acid	pK_{a1} DMSO	pK_{a1} H$_2$O	ΔpK_{a1}[b]	pK_{a2} DMSO	pK_{a2} H$_2$O	ΔpK_{a2}[b]
oxalic	6.2	1.23	5.0	14.9	4.29	10.6
malonic	7.2	2.87	4.5	18.6	5.67	12.9
succinic	9.5	4.20	5.3	16.7	5.55	11.1
glutaric	10.9	4.34	6.6	15.3	5.42	9.9
adipic	11.9	4.42	7.5	14.1	5.42	8.7
azelaic	12.0	4.55	7.4	13.6	5.41	8.2

[a] Ref. [18]; [b] $\Delta pK_{ai} = pK_{ai}(DMSO) - pK_{ai}(H_2O)$

increases ΔpK_{a2}; the result is a separation between the two steps of greater than 8 pK units. The behaviour of succinic acid ($n = 2$) and glutaric acid ($n =3$) is intermediate between these two extremes.

Chantooni and Kolthoff [18] have estimated log K' (eq. (6.5)) for malonic, succinic, glutaric and adipic acids by comparing the effect of solvent on the first dissociation of the dicarboxylic acids and their mono-esters (allowing for a statistical factor of 2 in K_{a1} because of the presence of two identical acid groups, compared with only one for the corresponding ester); the monoester anion is taken as a model for the non-hydrogen-bonded form of the monoanion. Thus we may equate K' to $K_{a1}(acid)^*/K_a(ester) - 1$, where $K_{a1}(acid)^* = K_{a1}(acid)/2$. The relevant data for the acids and their ethyl esters are as follows:

Acid[a]	oxalic	malonic	succinic	glutaric	adipic
$pK_{a1}(acid)^*$	6.5	7.5	9.8	11.2	12.2
$pK_a(ester)$	6.52	10.26	11.91	12.45[b]	12.64[b]
K'[c]	0	574	128	17	1.8

[a] $pK_{a1}(acid)^* = pK_{a1}(acid) + 0.30$ to allow for two identical acid groups;
[b] Methyl ester; [c] Eq. (6.5); $K' = K_{a1}(acid)^*/K_a(ester) - 1$

There is, as expected, no evidence for intramolecular H-bonding in the bioxalate ion, and K' is a maximum for malonic acid and decreases as the separation between the two carboxylic acid groups increases, becoming extremely small for adipic acid.

The same study reported that significantly larger values of K' occur in weakly basic aprotic solvents, e.g., acetonitrile, in which $K' = 2.5 \times 10^4$ for malonic acid. Larger effects in these latter solvents arise because they do not hydrogen bond significantly with either the carboxylic acid or the carboxylate groups, thus allowing maximum interaction between the carboxylic and carboxylate groups in the monoanion. IR-spectral evidence for intramolecular H-bonding in acetonitrile was also found.

Phenols and thiophenols

Dissociation constants for a variety of representative phenols [23, 24] are listed in Table 6.3 together with the corresponding values in water.

Table 6.3 pK_a-values of phenols in dimethylsulfoxide at 25°C[a]

Phenol	pK_a DMSO	$pK_a H_2O$	ΔpK_a[b]	Phenol	pK_a DMSO	$pK_a H_2O$	ΔpK_a[b]
2,4,6-trinitro	−0.7	0.43	−1.1	4-cyano	14.8	8.58	6.2
2,6-dinitro	5.4	4.10	1.3	3-chloro	15.8	9.02	6.8
3,5-dinitro	10.6	6.66	3.9	4-bromo	15.7	9.36	6.3
2-nitro	11.0	7.23	3.8	4-chloro	16.7	9.42	7.3
4-nitro	11.0	7.23	3.8	H	18.0	9.99	8.0
4-acetyl	14.1	8.02	6.1	4-methyl	18.9	10.26	8.6
3-nitro	14.4	8.36	6.0				

[a] Ref. [23, 24]; [b] $\Delta pK_a = pK_a(DMSO) - pK_a(H_2O)$

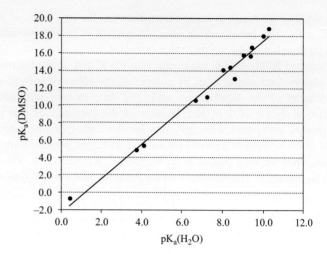

Fig. 6.2.
pK$_a$-values of phenols in dimethylsulfoxide versus water at 25°C

There is an excellent correlation between pK$_a$(DMSO) and pK$_a$(H$_2$O) for a wide range of phenols, shown in Fig. 6.2, with the best-fit line being given by eq. (6.6).

$$pK_a(DMSO) = 1.98 pK_a(H_2O) - 2.40 \qquad (6.6)$$

Particularly striking is the strong sensitivity to substituent of the solvent effect on pK$_a$, which means that the absolute changes in pK$_a$ on solvent transfer are very dependent upon the substrate involved. Thus phenol itself is weaker in DMSO by some 8.0 pK units, whereas picric acid, 2,4,6-trinitrophenol, *almost uniquely amongst neutral acids in non-aqueous solvents*, is slightly stronger in DMSO than in water.

The explanation for this lies predominantly in the influence of substituent on the charge density of the phenoxide anion, which in turn affects its solvation. As the negative charge on the phenoxide becomes increasingly dispersed, the importance of H-bond solvation in water is decreased, and the solvation of the anion is increasingly dominated by dispersion-force interactions, similarly to that of the neutral phenol; these are stronger with the highly polarizable DMSO molecules compared with water molecules. The change in free energy of highly charge-dispersed phenoxide ions, $\Delta G_{tr}(ArO^-)$, such as the picrate anion, thus approaches that of the corresponding phenol, $\Delta G_{tr}(ArOH)$. Under these conditions, it follows, from eq. (6.2), that ΔpK_a is then determined primarily by $\Delta G_{tr}(H^+)$, which is negative in DMSO because of its stronger Lewis basicity compared with water (Table 3.7).

Similar behaviour is exhibited by the thiophenols [25], with an excellent correlation between pK$_a$(DMSO) and pK$_a$(H$_2$O), represented by eq. (6.7), the data for which are shown in Table 6.4.

$$pK_a(DMSO) = 2.55 pK_a(H_2O) - 6.26 \qquad (6.7)$$

In quantitative terms it is apparent from a comparison of the results in Tables 6.3 and 6.4 that the thiophenols experience a significantly smaller

Table 6.4 pK_a-values of thiophenols in dimethylsulfoxide at 25°C[a]

Thiophenol	pK_a DMSO	$pK_a H_2O$	ΔpK_a[b]	Thiophenol	pK_a DMSO	$pK_a H_2O$	ΔpK_a[b]
4-nitro	5.60	4.72	0.82	3-methyl	10.55	6.60	3.95
2-chloro	8.55	5.68	2.87	H	10.28	6.62	3.66
3-chloro	8.57	5.78	2.79	4-methoxy	11.19	6.78	4.41
4-bromo	8.98	6.02	2.96				

[a] Ref. [25]; [b] $\Delta pK_a = pK_a(DMSO) - pK_a(H_2O)$

reduction in acidity (increase in pK_a) on transfer to DMSO compared with the corresponding phenols; for example, the increase in the pK_a of thiophenol of 3.8 units, is some 4.2 units smaller than that for phenol, 8.0 units. The difference resides almost entirely in the solvation of the two anions (Table 3.7) [26]: $\Delta G_{tr}(PhO^-)$ is some 25.9 kJ mol^{-1} larger than $\Delta G_{tr}(PhS^-)$, which equates to 4.5 pK-units. This reflects both the weaker solvation of the thiophenoxide ion in water by hydrogen bonding and the stronger dispersion force interactions with DMSO.

The increased sensitivity of the phenols and thiophenols to substituent effects relative to those in water, represented by the slopes in eqs. (6.6) and (6.7), 1.88 and 2.55, respectively, is more pronounced than for the carboxylic acids. This again is a consequence of the stronger variations with substituent in the charge densities of $-O^-$ and $-S^-$ relative to $-CO_2^-$, because of their direct conjugation to the aromatic ring.

Homohydrogen-bond constants for phenols, analogous to eq. (6.1), remain almost constant at $K_{AHA} = 2 \times 10^3$ M^{-1}, across a range of m- and p-substituted phenols, despite a large change in acidity [23]. Thus, with respect to homohydrogen-bond formation, an increased acidity, and hence a stronger tendency to donate an H-bond by the phenol, is counterbalanced by a decreased tendency of the correspondingly more weakly basic phenoxide ion to accept a H-bond. When measured against a common donor, however, the hydrogen-bond accepting capacity of ArO$^-$ increases with increasing basicity, as expected [27]. o-Substitution strongly reduces homohydrogen-bond formation.

There is no evidence of homohydrogen-bond formation between PhS$^-$ and PhSH in DMSO [25]. Evidently H-bonding of PhSH with DMSO is stronger than with PhS$^-$.

Water and methanol

The ionization of both water and methanol, eqs. (6.8) and (6.9), is strongly inhibited in DMSO, leading to large increases in their pK_a-values, and correspondingly greatly increased basicity of the hydroxide and methoxide ions [5, 14].

$$H_2O \rightleftharpoons H^+ + OH^- \quad K_a(H_2O) = [H^+][OH^-]/[H_2O] \tag{6.8}$$

$$MeOH \rightleftharpoons H^+ + MeO^- \quad K_a(MeOH) = [H^+][MeO^-]/[MeOH] \tag{6.9}$$

pK_a-values for water and methanol in DMSO are 32 and 29.0, respectively, compared with the corresponding values in the pure solvents of 15.75 and 18.15. The free energy of the hydroxide ion on transfer from water to DMSO has been estimated to increase by some 109 kJ mol^{-1} [28] (i.e., log$^{aq}\gamma^S$ = 19.1), and this is obviously the dominant factor in the very large increase in pK_a(H$_2$O).

The basicity of hydroxide and methoxide ions in water–DMSO and MeOH–DMSO mixtures, respectively, increases strongly with increasing DMSO content of the mixtures, and this has been used to advantage in promoting the ionization of weakly acidic substrates [29].

Anilines and amides

In aqueous solution there is little tendency of anilines to ionize, eq. (6.10)

$$ArNH_2 \xrightleftharpoons{pK_a(ArNH_2)} ArNH^- + H^+ \qquad (6.10)$$

Stewart and co-workers [30, 31] found, however, that hydroxide solutions in various DMSO–water mixtures are sufficiently basic to deprotonate them. For example, ionization of the 4-NO$_2$-derivative by 0.01M Bu$_4$NOH occurs at around 87 wt% DMSO and the 4-CN derivative at around 99 wt% DMSO. By means of an extrapolation procedure, based on the effective 'pH' of the various mixtures, H$_-$, they estimated aqueous pK_a-values for anilines acting as an acid; these ranged from 12.2 for 2,4,6-trinitro aniline to 27.7 for unsubstituted aniline.

Ionization of anilines can be readily achieved in DMSO by use of a suitably strong base, such as the dimsyl anion, and this has been used, in conjunction with the indicator method described earlier (Section 4.5.1) [32], to measure the dissociation constants for some 27 anilines. Representative values are listed in Table 6.5.

The anilines show similar sensitivity to substituents to that of the phenols, but are, on average, ∼ 12 pK units less acidic.

Aromatic amides, ArNHCOCH$_3$, are, as expected, somewhat more acidic than the corresponding anilines; acetanilide has a pK_a in DMSO of 21.5 compared with that of aniline of 30.7. They are also less sensitive to substituent effects (Section 6.5 below) [33].

Table 6.5 pK_a-values of anilines in dimethylsulfoxide at 25°C[a,b]

Aniline	pK_a(DMSO)	Aniline	pK_a(DMSO)
2,4-dinitro	15.9	3-trifluoromethyl	28.5
4-nitro	20.9	3-chloro	28.5
2,6-dichloro	24.8	4-bromo	29.1
4-acetyl	25.3	4-chloro	29.4
4-cyano	25.3	H	30.7
2,4-dichloro	26.3	3-methyl	31.0

[a] Dissociation of ArNH$_2$ to ArNH$^-$; [b] Ref [32]

Carbon acids

The defining characteristic of important classes of carbon acids, such as the ketones and nitroalkanes, is a structural rearrangement that accompanies the ionization of the C–H bond, such that the negative charge generated resides on oxygen rather than carbon, eqs. (6.11), (6.12).

$$\text{(ketone)} \underset{k_{-1}}{\overset{k_1}{\rightleftarrows}} \text{(enolate)} + H^+ : K_a = k_1/k_{-1} \qquad (6.11)$$

$$\text{(nitroalkane)} \underset{k_{-1}}{\overset{k_1}{\rightleftarrows}} \text{(nitronate)} + H^+ : K_a = k_1/k_{-1} \qquad (6.12)$$

The result is a considerable enhancement of the thermodynamic acidity of these carbon acids relative to those with an unsubstituted C–H bond. Aliphatic esters and the C–H bond of carboxylic acids ionize in the same way (Section 2.5). These acids are sometimes referred to as 'pseudo'-acids [34, 35], because of the structural rearrangement accompanying ionization, one consequence of which is that the *rate* of ionization is often orders of magnitude slower than that of comparable-strength 'normal' acids, i.e., acids in which the ionizing proton is bonded to an electronegative atom, such as oxygen or nitrogen.

In most other carbon acids, enhanced acidity results from the inductive influence of strongly electron-withdrawing groups α to the ionising C–H bond; examples of such groups include –SOR, –SO$_2$R, and –CN (Section 2.5).

In aqueous solution, the acidities of carbon acids are mostly very low, but it has in many cases proved possible to determine accurate pK_a-values through separate measurement of forward and reverse rate constants, k_1 and k_{-1}, eqs. (6.11) and (6.12), or indirectly through NMR-determinations of the rates of H/D exchange [36, 37]. The latter are controlled by the rate of C–H ionization (k_1), which can be combined with estimates of the rate constants for the (rapid) reverse reactions, k_{-1}, to give K_a.

In DMSO, direct measurement of the dissociation constants is possible by applying the indicator method to known ratios of the acid and conjugate base generated by addition of strong bases, such as the potassium or caesium dimsyl salts (Section 4.5.2) [5]. Representative values of the dissociation constants are listed in Table 6.6.

The mono-substituted carbon acids in which the negative charge resides on an oxygen atom, such as ketones and nitroalkanes, mostly typically show pK-increases of around 6.5 units compared to their aqueous values, which are similar to those of the weaker phenols and carboxylic acids (Tables 6.1, and 6.3). This shows that solvation of the anions in water has an important influence on the aqueous acidity of these acids. By contrast, HCN and CH$_3$CN show much smaller increases in pK_a, consistent with weaker solvation of the CN$^-$ and CH$_2$CN$^-$ ions in water. Di-substituted acids normally have acidities in DMSO that are much closer to those in water, and in the case of malonitrile

Table 6.6 pK_a-values of carbon acids in dimethylsulfoxide at 25°C[a]

Acid	pK_a DMSO	$pK_a H_2O$	ΔpK_a[b]	Acid	pK_a DMSO	$pK_a H_2O$	ΔpK_a[b]
CH_3COCH_3	26.5	19.3	7.2	$(CH_3CO)_2CH_2$	13.3	9.1	4.2
$PhCOCH_3$	24.7	18.2	6.5	$EtCO_2CH_2NO_2$	9.1[b]	5.8	3.3
CH_3NO_2	17.2	10.2	7.0	$(NO_2)_2CH_2$	6.6	5.2	1.4
$CH_3CH_2NO_2$	16.4	8.8	6.6	$(MeSO_2)_2CH_2$	15.0	12.7	2.3
CH_3CO_2Et	29.5	25.6	3.9	$(EtSO_2)_2CHCH_3$	16.7	14.5	2.2
CH_3SOCH_3	35.1	33	2.1	$EtCO_2CH_2CN$	12.5[c]	10.2	2.3
CH_3CN	31.3	28.9	2.4	$(CN)_2CH_2$	11.1	11.2	−0.1
HCN	12.9	9.2	3.7	$(CN)_2CHPh$	4.2		

[a] Ref. [5]; Bordwell, F.G.; Harrelson, J.A. *Can. J. Chem.*, 1990, 68, 1714; [b] $\Delta pK_a = pK_a(DMSO) - pK_a(H_2O)$; [c] Goumont, R.; Magnier, E.; Kizilian, E.; Terrier, F. *J. Org. Chem.*, 2003, 68, 6566

*Similar protolytic equilibria occur for the nitroalkanes, where protonation of the nitroalkane anion can occur on either the carbon atom or the oxygen atoms to give the *aci*-form; in common with enolate equilibria, oxygen protonation is kinetically favoured, but carbon protonation is the thermodynamically favoured process [34]

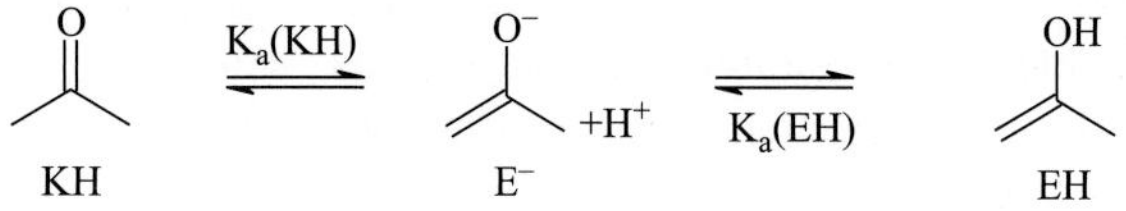

slightly lower than in water, due to the more strongly dispersed negative charge of the anions.

There are two ionization processes associated with ketones: the dissociation of the ketone and the dissociation of the corresponding enol, with the equilibrium constant for enol formation being given by the ratio of the two dissociation constants, as in Scheme 6.2 for acetone.*

$$K_E = [EH]/[KH] = K_a(KH)/K_a(EH)$$

Solvent	$pK_a(KH)$	$pK_a(EH)$	pK_E
Dimethylsulfoxide[a]	26.5	18.2	8.3
Water[b]	19.3	10.9	8.4

[a] Ref [5]; Bordwell, F.G.; Zhang, S.; Eventova, I.; Rappoport, Z. *J. Org. Chem.*, 1997, 62, 5371; [b] Kresge, A.J.; Tobin, J.B. *J. Am. Chem. Soc.*, 1990, 112, 2805

Scheme 6.2. Keto-enol equilibria of acetone

It can be seen from the data for $K_a(KH)$ and $K_a(EH)$ in Scheme 6.2 that the individual acidity constants for the ketone and enol decrease by an almost exactly equal amount (7.3 and 7.2 pK-units, respectively) on transfer from water, both effects being dominated by the decreased solvation of the common enolate anion. The result is that the equilibrium proportion of the enol remains constant at less than 1 part in 10^8 in the two solvents.

Dissociation constants of additional series of carbon acids in DMSO are presented below in Section 6.5 (Table 6.14).

6.1.2 Cationic acids (neutral bases)

Table 6.7 lists the dissociation constants for various (protonated) anilines, amines and pyridine in DMSO [38–42], together with their values in water for comparison.

The broad conclusion from these results is that the differences between water and DMSO are small, but in almost all cases the protonated nitrogen bases are more acidic in DMSO. This is consistent with a stronger solvation

Table 6.7 pK_a-values of anilinium, ammonium, and pyridinium ions in dimethylsulfoxide at 25°C

Amine[a]	pK_a DMSO	$pK_a H_2O$	ΔpK_a[b]	Amine/aniline	pK_a DMSO	$pK_a H_2O$	ΔpK_a[b]
ammonia	10.5	9.21	1.3	tri-*n*-butylamine	8.4	10.9	−2.5
methylamine	11.0	10.65	0.3				
ethylamine	10.7	10.67	0	3-cyanoaniline[c]	1.36	2.75	−1.4
n-propylamine	10.7	10.57	0.1	3-chloroaniline[c]	2.34	3.46	−1.1
n-butylamine	11.12	10.59	−0.5	4-chloroaniline[c]	2.86	3.98	−1.1
dimethylamine	10.3	10.78	−0.5	aniline[c,d]	3.82	4.85	−1.0
diethylamine	10.5	11.0	−0.5	*N*-Me-aniline[d]	2.76	4.85	−2.1
di-*n*-butylamine	10.0	11.3	−1.3	*N*,*N*-Me$_2$-aniline[d]	2.51	5.16	−2.7
piperidine	10.85	11.12	−0.3				
pyrrolidine	11.06	11.31	−0.2	pyridine[d]	3.4	5.22	−1.8
trimethylamine	8.4	9.80	−1.4	Me$_4$-guanidine	13.2	13.6	−0.4
triethylamine	9.0	10.67	−1.7				

[a] Ref. [38]; Ref. [40, 41]; [b] $\Delta pK_a = pK_a(DMSO) - pK_a(H_2O)$; [c] Ref. [39]; [d] Ref. [42]

of both the proton and the free base in DMSO, increasing the tendency to dissociate, which will be counterbalanced to some extent by stronger solvation of the corresponding ammonium cation.

Closer examination of the results in Table 6.7 reveals also that that within a related series of bases there is a systematic difference between the behaviour of primary, secondary and tertiary amines. The largest increases in acidity occur for the most highly substituted nitrogen bases, and the increases in acidities show a general trend of tertiary amine (∼ 1.9 units) > secondary amine (∼ 0.8 units) > primary amine (∼ 0) > NH_4^+ (−1.3 units). Similarly, the increase in acidity of the anilines follows the order *N*,*N*-dimethylaniline (2.7 units) > *N*-methylaniline (2.1 units) > aniline (1.1 units). Most probably these trends result from increasingly strong interactions between the aminium cations and DMSO as the number of N−H protons increases, because of the high Donor Number and hydrogen-bonded basicity of DMSO. This will have the effect of reducing the acidity of NH_4^+ in DMSO relative to methylamine, dimethylamine, etc. Such a distinction disappears in weakly basic, aprotic solvents, such as acetonitrile (Chapter 7).

The trend in the acidities of the various cations in DMSO with respect to *N*-alkyl-substitution is opposite to that observed in the gas-phase. In the latter case, the basic strengths (proton affinities) increase directly with increasing alkyl substitution [43], as the primary requirement in the gas phase is for stabilization of the cationic charge; the larger the alkyl groups, the more effective this is.

6.1.3 Amino acids

Amino acids and peptides are amongst the most widely studied classes of compounds in chemistry. They are frequently used as sources of enantiomeric purity in organic syntheses, and much is known about their physical properties and physical constants in aqueous solution. In enzymes and *in vivo* reactions, however, the medium is often lipophilic rather than hydrophilic, and hence their pK_a-values are likely to be very different from those in water [44]. We have earlier noted that for amino-benzoic acids in ethanol–water mixtures, the extent of zwitter-ion formation decreases strongly with increasing ethanol content of the solvent (Section 5.5).

The relationship between the equilibrium portion of zwitter-ion and the possible modes of dissociation of protonated amino acids is illustrated in Scheme 6.3 for glycine.

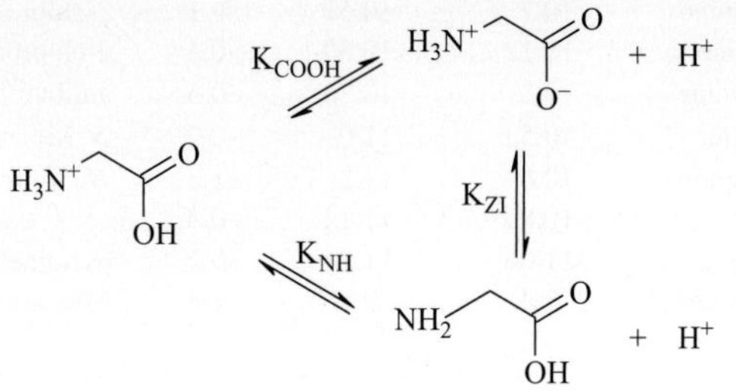

Scheme 6.3.
The dissociation of protonated glycine

It follows from Scheme 6.3 that the zwitter-ion constant, $K_{ZI} = $ [zwitter-ion]/[neutralform] is related to the dissociation constants for the alternative dissociation processes by eqs. (6.13) and (6.14).

$$K_{ZI} = K_{COOH}/K_{NH} \tag{6.13}$$

$$\log K_{ZI} = pK_a(NH) - pK_a(COOH) \tag{6.14}$$

In the majority of cases experimental methods for the determination of dissociation constants do not distinguish between the zwitter-ion and neutral forms, so that the observed dissociation constant for the protonated amino acid, $K_{a1}(glyH^+)$, corresponds to eq. (6.15), in which ZI represents the zwitter-ion and NF the neutral form.

$$K_{a1}(glyH^+) = \{[ZI] + [NF]\}[H^+]/[glyH^+] \tag{6.15}$$

In aqueous solution the proportion of the neutral form of aliphatic amino acids is very low and is difficult to measure directly, because of the much higher dissociation constants of carboxylic acids compared with protonated aliphatic amines. Under these circumstances, $K_{COOH} = K_{a1}$ to a very good approximation, and it is therefore difficult to determine K_{NH} directly; rather it is most commonly estimated by approximating it to K_a for the methyl ester of glycine, eq. (6.16) ($K_a = K_{ENH}$).

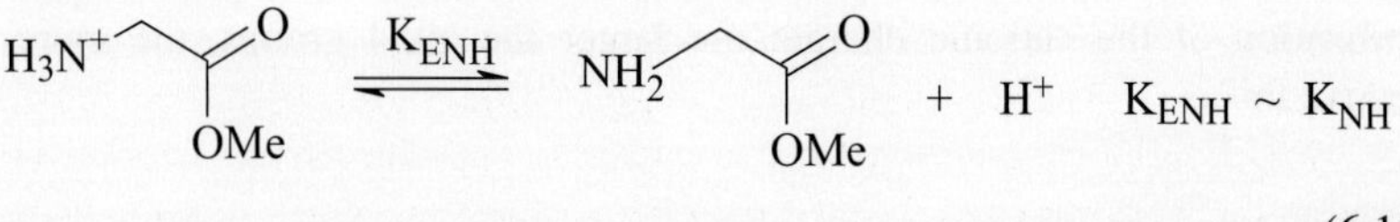

$$\tag{6.16}$$

Substituting the resultant pK_a-values for K_{COOH} and K_{NH}, 2.43 and 7.66, respectively into eq. (6.15) gives $K_{ZI} = 1.7 \times 10^5$ (log $K_{ZI} = 5.2$). On the basis of the effect of DMSO on the dissociation constants of carboxylic acids and primary amines reported above, i.e., increases of ~ 5 and 0 units, respectively (Tables 6.1 and 6.7), K_{ZI} would be expected to decrease strongly towards unity in DMSO.

Table 6.8 Dissociation constants of amino acids in dimethylsulfoxide at 25°C[a]

Amino acid	pK_{a1}[b]	pK_{COOH}[c]	pK_{NH}[c]	K_{ZI}[c]
$NH_3^+CH_2COOH$	7.5	7.5	9.1	40
$CH_3NH_2^+CH_2COOH$	6.8	6.8	8.25	27
$(CH_3)_2NH^+CH_2COOH$	6.3	6.45	6.78	2

[a] Ref. [44]; [b] First dissociation constant, eq. (6.15); [c] Scheme 6.1

Table 6.8 lists relevant dissociation constants and the resulting K_{ZI} for glycine derivatives in DMSO [44].

The low value of K_{ZI} for N,N-dimethylglycine means that the zwitter-ion and neutral forms coexist in comparable quantities, allowing the approximation that $K_{NH} \sim K_{ENH}$, eq. (6.16) to be tested by direct measurement of their ratio in DMSO using ^{1}H-NMR. It was found that pK_{NH} was approximately 0.4 units lower than that for the ester, pK_{ENH}, eq. (6.16), and this difference was assumed to hold also for glycine and N-methylglycine in calculating the K_{NH} reported for them in Table 6.8.

The systematic decrease in K_{ZI} on going from the primary to secondary to tertiary amine is consistent with a correspondingly reduced basicity of the amino group, as observed for simple amines, Table 6.7.

6.2 *N*-methylpyrrolidin-2-one, *N, N*-dimethylformamide, *N, N*-dimethylacetamide

6.2.1 Neutral acids

Data in these solvents are less comprehensive than in DMSO, but are nevertheless sufficient to establish useful trends and correlations. In view of the comprehensive set of results available in DMSO, including importantly a wide range of carbon acids, we begin by comparing the results for these solvents with those in DMSO.

An examination of the free energies of transfer of ions in the different solvents (Table 3.7) suggests that trends in acidity with structure and substituent should be very similar to those observed in DMSO, but with absolute values of the dissociation constants that are slightly lower than those in DMSO. Thus, $\Delta G_{tr}(H^+ + OAc^-)$ in NMP and DMF are, respectively, 48.2 and 50.1 kJ mol^{-1} (or equivalently, 8.5 and 8.8 in terms of log $^{aq}\gamma^S$), compared with 41.7 kJ mol^{-1} (log$^{aq}\gamma^S = 7.3$) in DMSO. This should correspond to increases in pK_a of some 1.2 and 1.5 units in NMP and DMF, respectively, compared with DMSO, given the small difference between the solvation of acetic acid (and other neutral acids) in the three solvents.

Starting with NMP, there is indeed an extremely good correlation between the pK_a-values for neutral acids with those in DMSO, *irrespective of the nature of the acid*. Fig. 6.3 shows a comparison of pK_a-values for a range of neutral acids in NMP and DMSO, covering some 30 pK-units, reported by Bordwell

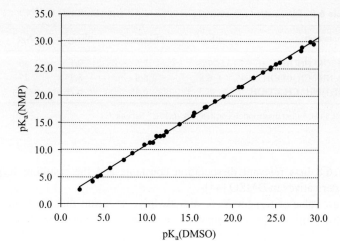

Fig. 6.3.
Comparison of pK_a-values of neutral acids in dimethylsulfoxide and N-methylpyrrolidin-2-one at 25°C

and co-workers [45]. The acids include naphthols, phenols, carboxylic acids, fluorenes, anilines, nitriles, and sulphones. There is an almost constant difference of 1.1 units between pK_a-values in the two solvents, in agreement with the discussion above. The best-fit line, included in Fig. 6.3, corresponds to eq. (6.17); the slope is notably very close to unity.

$$\mathrm{p}K_a(\mathrm{NMP}) = 0.99\mathrm{p}K_a(\mathrm{DMSO}) + 1.08 \qquad (6.17)$$

Dissociation constants of representative acids in NMP are listed in Table 6.9.

All are covered by the relationship in eq (6.17), except for phenol, 3-chlorophenol and benzoic acid, for which the pK_a-values are almost 2 units higher in NMP than in DMSO. The excellence of the correlation in Fig. 6.3 suggests that the outliers in the results may be due at least in part to experimental uncertainties, although the values in Table 6.9 are considered to be the most reliable available. Difficulties in measurement can arise because of

Table 6.9 pK_a-values of neutral acids in N-methylpyrrolidin-2-one at 25°C[a]

Acid	pK_a(NMP)	Acid	pK_a(NMP)
2,4-$(NO_2)_2$-1-naphthol	2.72	3-Cl-phenol	17.8
4-Cl-2,6-$(NO_2)_2$-phenol	4.22	$(EtSO_2)_2CHCH_3$	17.9
$PhCH(CN)_2$	5.1	9-phenylfluorene	19.05
F_3CSO_3NHPh	6.7	4-Cl-2-NO_2-aniline	19.95
9-CN-fluorene	9.45	phenol	20.1
4-NO_2-phenol	12.15	9-Me-fluorene	23.4
CH_2CN_2	12.6	Ph_2NH	25.9
benzoic acid	13.3	$CH_3SO_2CH_2Ph$	26.2
acetic acid	13.6[b]	$p-ClC_6H_4SO_2CH_3$	29.0
$(PhSO_2)_2CH_2$	13.45	$PhSO_2CH_3$	30.0
4-acetylphenol	15.6	Ph_3CH	31.0

[a] Ref. [45]; [b] Calculated from eq. (3.4), using ΔG_{tr} values from Table 3.7 and 3.8

the large homohydrogen-bond constants in both solvents, and electrochemical measurements can also be influenced by a sluggish electrode response at high pH-values [16]; indeed, in extreme cases literature values reported for these acids in different laboratories can vary by as much as five units.

A similarly excellent correlation between pK_a for neutral acids in dimethylformamide and dimethylsulphoxide also exists, with results for some 84 acids being represented by eq. (6.18) [46].

$$pK_a(DMF) = 0.96 pK_a(DMSO) + 1.56 \qquad (6.18)$$

The acids included in the correlation include benzoic acids, phenols, mono- and dicarboxylic acids, sulphonamides, amides and NH-heterocycles. Measurements have been mostly performed using standard potentiometric techniques, apart from the NH-heterocycles, which were determined using voltametric techniques anchored to results from more traditional methods using overlapping acids. Additional data for a variety of substituted benzoic acids are provided by Pytela and co-workers [19, 20, 47, 48].

Representative dissociation constants in DMF are listed in Table 6.10.

Dicarboxylic acids in DMF show very similar behaviour to that in DMSO (Table 6.2), with strong evidence for intramolecular H-bond stabilization of the mono-carboxylate anion in the case of malonic and succinic acids; individual pK_a-values are typically around one unit higher than in DMSO for the first dissociation and two units higher for the second dissociation [46, 49].

Dimethylacetamide has polarity and solvation properties closest to those of NMP (Tables 1.1 and 3.7) but similar also to those of DMF and DMSO. In practice, the available dissociation constants (Table 6.11) are very close to those in DMSO [47].

Among the dicarboxylic acids, the difference between the first and second dissociation constants decreases sharply from malonic acid ($\Delta pK = 9.6$) to adipic acid ($\Delta pK = 1.2$), consistent with a strong intramolecular H-bond

Table 6.10 pK_a-values of neutral acids in N,N-dimethylformamide at 25°C[a]

Acid	pK_a(DMF)	Acid	pK_a(DMF)
dichloroacetic	7.6	4-NO_2-phenol	12.3
2-chloroacetic	10.2	3-NO_2-phenol	14.6
acetic	13.5	3-CF_3-phenol	15.7
2,4-$(NO_2)_2$-benzoic	8.2	3-Cl-phenol	16.3
3,5-$(NO_2)_2$-benzoic	8.8	4-Cl-phenol	16.8
3,5-$(Cl)_2$-benzoic	10.4	phenol	> 18[b]
4-NO_2-benzoic	10.6	4-NO_2-thiophenol	6.3
3-NO_2-benzoic	10.7	thiophenol	10.7
3-Br-benzoic	11.2	formanilide	20.3
4-Cl-benzoic	11.5	acetanilide	22.3
benzoic	12.3	nicotinamide	22.5
4-NH_2-benzoic	14.0	benzamide	23.9
3,5-$(NO_2)_2$-phenol	11.3	Ph_2NH	25.5

[a] Ref. [46]; [b] The correlation represented by eq. (6.18) suggests a value of 18.9

Table 6.11 pK$_a$-values of neutral acids in N, N-dimethylacetamide at 25°C[a]

Acid	pK$_a$(DMAC)	Acid		pK$_a$(DAMC)
dichloroacetic	4.5	malonic	pK$_{a1}$	6.5
2-chloroacetic	8.8		pK$_{a2}$	16.1
lactic	9.9	succinic	pK$_{a1}$	8.9
acetic	12.6		pK$_{a2}$	14.6
2-chlorobenzoic	9.6	glutaric	pK$_{a1}$	10.2
3-chlorobenzoic	9.8		pK$_{a2}$	13.0
4-chlorobenzoic	10.3	adipic	pK$_{a1}$	11.2
benzoic	11		pK$_{a2}$	12.4

[a] Ref. [50]

stabilization on the mono-carboxylate in the former (see Section 6.1.1); succinic and glutaric acids show intermediate behaviour.

The influence of intramolecular stabilization by hydrogen-bonding is particularly apparent in a comparison of the dissociation constants of the isomeric, unsaturated acids, maleic and fumaric acids in DMAC, illustrated in Scheme 6.4, in which $\Delta pK = pK_{a2} - pK_{a1}$ [50].

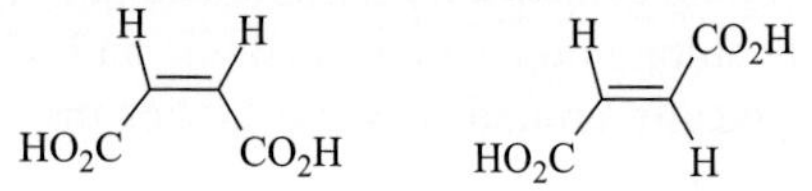

	maleic	fumaric
pK$_{a1}$	4.1	9.3
pK$_{a2}$	17	11.6
Δ pK	12.9	2.3

Scheme 6.4.
Dissociation constants of maleic and fumaric acids in dimethylacetamide [50]

The contrast in behaviour of the two acids is stark. Although the combined pK$_a$(= pK$_{a1}$ + pK$_{a2}$) for the complete dissociation into the dicarboxylate anions is closely similar for the two acids, 21.1 and 20.9, respectively, the intermediate mono-carboxylate is clearly very strongly stabilized in maleic acid compared with fumaric acid, resulting in a substantially lower pK$_{a1}$ and a correspondingly higher pK$_{a2}$. Very similar behaviour for these acids is also observed in NMP [51].

6.2.2 Cationic acids (neutral bases)

A relatively limited set of data for protonated amines and related cationic acids is available in DMF [52–54], NMP [51, 55], and DMAC [51]. Dissociation constants are listed in Table 6.12, along with those in water and DMSO for comparison.

Primary amines in each of the solvents are normally similar to or slightly less acidic than in water, whereas secondary and tertiary amines are somewhat more acidic, as discussed in Section 6.1.2. Among the aprotic solvents, pK$_a$-values in DMF are very similar to those in DMSO, whereas those in NMP and DMAC are on average about 0.6 units lower.

Table 6.12 pK$_a$-values of ammonium ions in basic, aprotic solvents at 25°C

Amine	pK$_a$H$_2$O	pK$_a$DMSO[a]	pK$_a$DMF[b]	pK$_a$NMP[c]	pK$_a$DMAC[d]
ammonia	9.21	10.5	9.45		
ethylamine	10.67	10.7	10.2		
n-propylamine	10.57	10.7			
n-butylamine	10.59	10.9	10.3		
dimethylamine	10.78	10.3	10.4	9.1	
diethylamine	11.0	10.5	10.4	9.3	9.1
triethylamine	10.67	9.0	9.25	8.8	
tri-n-butylamine	10.9	8.4	8.57		
triethanolamine	7.5	7.6	7.3	6.8	
pyridine	5.22	3.4	3.3		
Me$_4$-guanidine	13.6	13.2	13.65	12.9	

[a] Table 6.7; [b] Ref. [52–54]; [c] Ref. [51, 55]; [d] Ref. [51]

6.3 Liquid ammonia

Liquid ammonia acts as a typical, but strongly basic, polar aprotic solvent [56]. It has promise as a possible replacement in industrial processes for more commonly used polar aprotic solvents, such as DMSO, DMF, and NMP, because of its cheapness and relative ease of recycling.

The relevant properties of liquid NH$_3$ may be summarized as follows [57]. It has a boiling point of −33°C and a vapour pressure of 8 bar at 25°C. The nitrogen lone pair makes it a very good hydrogen-bond acceptor and a good solvent for cations, as reflected in the ^{23}Na-NMR chemical shifts [58] and a Donor Number, DN = 58.2, almost twice that of DMSO. Unlike water and other protic solvents, however, it is a very poor hydrogen-bond donor and does not significantly solvate anions [59, 60].

The dissociation of acids, HA, may be represented by eq. (6.19), in which we have specifically included the ammonia in the equilibrium because of the obvious residence of the proton on an NH$_3$ molecule.

$$HA + NH_3 \xrightleftharpoons{K_a} NH_4^+ + A^- \xrightleftharpoons{K_{IP}} [NH_4^+ A^-] \quad (6.19)$$

The modest dielectric constant, $\varepsilon_r = 16.9$, means that ion-pair formation will, in general, be a significant factor except in solutions of very low concentration.

The ionization of phenols generates anions with a convenient UV absorption that can be used to determine the overall ionization equilibria, which includes formation of the free (K$_a$) and ion-paired (K$_{IP}$) phenoxide ions [62]. The two equilibria were separated either by extrapolation to zero ionic strength or by measuring the effect of added NH$_4^+$, and the resulting dissociation constants are given in Table 6.13. Phenols with aqueous pK$_a$ < 7 in water are fully ionized in liquid ammonia at room temperature, but not those with pK$_a$ > 8.5.

It is instructive in terms of the influence of solvent basicity on the dissociation constants in aprotic solvents to compare the dissociation constants of the phenols in liquid ammonia (strongly basic), dimethylsulphoxide (basic) and acetonitrile (weakly basic, Chapter 7), with those in water, as in Fig. 6.4

Table 6.13 pK_a-values of phenols in liquid ammonia at 25°C[a,b]

Phenol	pK_aNH_3	pK_aH_2O
4-nitrophenol	1.10	7.14
3-nitrophenol	3.61	8.36
4-carbomethoxyphenol	4.04	8.47
3-chlorophenol	4.50	9.02
4-chlorophenol	4.69	9.20
1-naphthol	4.97	9.37
phenol	6.02	9.99
4-methoxyphenol	6.62	10.27

[a] Ref. [62]; [b] Zero ionic strength

The slopes of the plots, 1.86 (MeCN), 1.97 (DMSO) and 1.68 (liquid NH_3) are, as expected, all significantly larger than 1, reflecting the poor anion solvation in these aprotic solvents, but the most striking feature is the strong increase in dissociation constant with increased solvent basicity; the pK_a of phenol decrease from 28.5 in MeCN to 18.0 in DMSO and to 6.02 in liquid NH_3, *an overall decrease of more than 22 pK-units*.

The ionization of carbon acids in liquid ammonia has also been reported recently [62]. Carbon acids with an aqueous pK_a of less than 11, such as HCN ($pK_a = 9.2$), acetylacetone ($pK_a = 9.0$), and malononitrile ($pK_a = 11.2$), are full ionized in liquid ammonia. Weaker acids have pK_a-values which are considerably lower than those either in water or DMSO: for example acetone, $pK_a(NH_3) = 16.5$, ethyl acetate, $pK_a(NH_3) = 18.2$, and acetonitrile, $pK_a(NH_3) = 18.3$; compare with DMSO pK_a-values of 26.5, 29.5 and 31.3, respectively. Again, the reduced free energy of the proton in strongly-basic liquid ammonia is a dominant factor in the increased acidities of these carbon acids.

The dissociation of aminium ions in liquid ammonia was studied by ^{1}H-NMR at 25°C using, as an indicator, the chemical shift difference of the protonated and free base forms of the amine seen in other solvents [61]. Trifluoroethylamine hydrochloride (aqueous $pK_a = 5.8$), is, as expected, fully

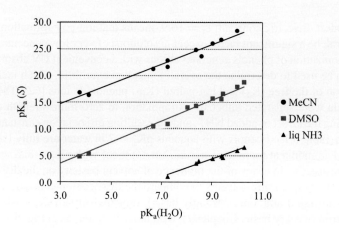

Fig. 6.4.
Comparison of pK_a-values of phenols in acetonitrile, dimethylsulfoxide, and liquid ammonia with those in water at 25°C

deprotonated in liquid ammonia, but surprisingly the hydrochlorides of more basic amines, such as piperidine (aqueous $pK_a = 11.27$) were also converted quantitatively to their free base forms. The equilibrium, eq. (6.20), must lie well over to the right, suggesting that ammonia solvent stabilizes the ammonium ion (NH_4^+) more than the aminium ions (RNH_3^+).

$$RNH_3^+ + NH_3 \xrightleftharpoons{K_e} NH_4^+ + RNH_2 \quad (6.20)$$

6.4 Summary

- The most remarkable feature of the results for neutral acids in these basic, aprotic solvents is the stark contrast between their transfer among the aprotic solvents and their transfer from water to any of the various solvents. For a wide variety of structurally distinct acids, including carboxylic acids, phenols, thiophenols, anilines, anilides, ketones, nitroalkanes, sulphones, and nitriles, *the change in pK_a for a neutral acid between any given pair of aprotic solvents is essentially constant and independent of the acid.* Conversely, transfer from any of the solvents to water depends critically upon the nature of the acid; for example carboxylic acids versus phenols, eqs. (6.2) and (6.6). Thus, in the aprotic solvents, the difference between the solvation of the neutral acid and its conjugate anion *does not depend upon the specific nature of the anion, the extent of dispersion of the negative charge, or whether the charge resides primarily on oxygen, sulphur, nitrogen, or carbon.* In water, however, the dominance of anion solvation via hydrogen-bond formation leads to a strong discrimination between the different types of anion in terms of their solvation.
- Linear correlations between the pK_a-values of neutral acids in DMSO and the other basic, polar aprotic solvents may be expressed by eq. (6.21) .

$$pK_a(S) = m pK_a(DMSO) + c \quad (6.21)$$

The best-fit values of m and c for the different solvents are:

Solvent, S	m	c
NMP	0.99	1.08
DMF	0.96	1.56
DMAC	1.0	0.1

In applying eq. (6.21), it should be noted that whereas the correlations in NMP and DMF are very well established, that in DMAC is more tentative.

- The neutral acids exhibit pK_a-values that are substantially higher than those in water, up to 8 pK-units in DMSO and correspondingly higher in DMF and NMP;* they are also considerably more sensitive to substituent effects.
- Cationic acids are generally slightly stronger in the basic aprotic solvents compared with water, the effects being largest for protonated tertiary amines.

*The only exception are acids, such as picric acid, whose conjugate bases have highly dispersed charges

Table 6.14 Hammett ρ-values for equilibrium acidities in dimethylsulfoxide at 25°C[a]

Acid family	$pK_a^{o\ b}$	ρ	Acid family	$pK_a^{o\ b}$	ρ
$ArCH(CN)_2$	4.2	4.2	Fluorenes[c]	22.6	7.5
$ArSO_2H$	7.1	2.4	Fluorenes[c]	22.6	5.7
ArSH	10.2	4.8	Phenolthiazines	22.7	5.21
$ArCO_2H$	11.0	2.6	$ArCH_2SO_2Ph$	23.4	4.8
ArCONHOH	13.65	2.6	$ArCOCH_3$	24.7	3.55
ArOH	18.0	5.3	ArNHPh	25.0	5.4
$ArCH_2COCH_3$	19.0	4.7	GCH_2CONH_2	25.5	3.1
$ArNHCOCH_3$	21.45	4.1	$ArNH_2$	30.6	5.7
$ArCH_2CN$	21.9	5.9	$ArCHPh_2$	30.6	5.7
$ArCH(C_6H_4O)CN$	22.4	7.0			

[a] Ref. [5]; [b] pK_a of parent (unsubstituted) acid; [b] 2- and 2,7-substituents; [c] 3-substituents

6.5 Estimation of dissociation constants in basic aprotic solvents

There are several ways in which to approach the problem of estimating dissociation constants in the basic, aprotic solvents:[†] via established correlations for transfer from water to aprotic solvents; via measured values in related aprotic solvents; or via existing data for structurally related acids in the same solvent.

The most useful starting point for neutral acids is the extensive set of dissociation constants in DMSO, and for simple acids and bases, correlations with corresponding data in water, Figs 6.1, 6.2, eqs. (6.2), (6.6), and (6.7), can be used to supplement the data. An alternative, particularly apposite for weakly acidic substrates, such as the various carbon acids (Section 6.1), is to use linear free energy relationships within DMSO as a basis for estimation of pK_a'-values of related substrates.

An example is the Hammett equation relating the pK_a of a substituted acid to that of the (unsubstituted) reference acid, pK_a^o, eq. (6.22).

$$pK_a = pK_a^o - \rho(\Sigma\sigma) \qquad (6.22)$$

In eq. (6.22), σ is a constant assigned to an individual substituent, and ρ is a constant for a particular acid–base system.

Similar additive free energy changes are observed for aliphatic acids, expressed by the Taft equation, eq (6.23), in which ρ* and σ* are the corresponding constants for aliphatic systems.

$$pK_a = pK_a^o - \rho^*(\Sigma\sigma^*) \qquad (6.23)$$

Extensive tabulations of ρ/ρ* and σ/σ* values are given by Perrin, Dempsey, and Serjeant [63], and Exner [64].

Hammett ρ-values for equilibrium acidities in DMSO are given in Table 6.14. The data for the most part is restricted to meta-substitution; full details are given by Bordwell [5].

For the remaining basic, aprotic solvents, comparisons with aqueous data are more difficult because of the relatively limited sets of data are available for

[†] The methods apply equally to the lower polarity and more weakly basic aprotic solvents, Chapter 7

the different classes of substrate. More useful in this context are the established correlations between DMSO and the other aprotic solvents, Section 6.3, which are independent of the nature of the acid, and hence can provide access to a wide range of neutral acids.

Cationic acids show dissociation constants that differ little from their aqueous values, except in the strongly basic liquid ammonia, but the changes do show specific differences between primary, secondary and tertiary amines (Sections 6.1.2 and 6.2.2).

References

[1] Gutmann, V. 'The Donor–Acceptor Approach to Molecular Interactions', Plenum Press, 1978
[2] Ehrlich, R. H., Roach, E., Popov, A. I. *J. Am. Chem. Soc.,* 1970, *92,* 4989
[3] Marcus, Y. *J. Soln. Chem.,* 1984, *13,* 599
[4] Abraham, M. H., Doherty, R. M., Kamlet, M. J., Taft, R. W. *Chem. Britain,* 1986, 551
[5] Bordwell, F. G. *Acc. Chem. Res.,* 1988, *21,* 456, and references therein
[6] Chantooni, M. K., Kolthoff, I. M. *J. Phys. Chem.,* 1976, *80,* 1306
[7] Ritchie, C. D., Uschold, R. E. *J. Am. Chem. Soc.,* 1968, *90,* 2821
[8] Lee, Ki P., Trochimowicz, H. J. *Journal of the National Cancer Institute,* 1982, *68,* 157
[9] Kaljurand, I., Kütt, A., Soovāli, L., Rodima, T., Mäemets, V., Leito, I., Koppel, I. A. *J. Org. Chem.,* 2005, *70,* 1019
[10] Kolthoff, I. M., Chantooni, M. K. *J. Chem. Eng. Data,* 1999, *44,* 124
[11] Coetzee, J. F, Padmanabhan, G. R. *J. Am. Chem. Soc.,* 1965, *87,* 5005
[12] Espinosa, E., Bosch, E., Rosés, M. *J. Chromatog. A,* 2002, *964,* 55
[13] Olmstead, W. N., Margolin, Z., Bordwell, F. G. *J. Org. Chem.,* 1980, *45,* 3295
[14] Kolthoff, I. M. Chantooni, M. K. *J. Phys. Chem.,* 1968, *72,* 2270
[15] Izutsu, K. 'Acid–Base Dissociation Constants in Dipolar Aprotic Solvents' IUPAC Chemical Data Series No. 35, Blackwell, 1990
[16] Matthews, W. S., Bares, J. E., Bartmess, J. E., Bordwell, F. G., Cornforth, F. J., Drucker, G. E., Margolin, Z., McCallum, R. J., Mcollum, J., Vanier, N. R. *J. Am. Chem. Soc.,* 1975, *97,* 7006
[17] Kolthoff, I. M., Chantooni, M. K. *J. Am. Chem. Soc.,* 1971, *93,* 3843
[18] Chantooni, M. K., Kolthoff, I. M. *J. Phys. Chem.,* 1975, *79,* 1176
[19] Pytela, O., Kulhánek, J. *Collect. Czech. Chem. Commun.,* 1997, *62,* 913
[20] Pytela, O., Kulhánek, J., Ludwig, M., Říha, V. *Collect. Czech. Chem. Commun.,* 1994, *59,* 627
[21] Chantooni, M. K., Kolthoff, I. M. *J. Phys. Chem.,* 1973, *77,* 527
[22] Chantooni, M. K., Kolthoff, I. M. *J. Phys. Chem.,* 1974, *78,* 839
[23] Bordwell, F. G., McCallum, R. J., Olmstead, W. N. *J. Org. Chem.,* 1984, *49,* 1424
[24] Chantooni, M. K., Kolthoff, I. M. *J. Phys. Chem.,* 1976, *80,* 1306
[25] Bordwell, F. G., Hughes, D. L. *J. Org. Chem.,* 1982, *47,* 3224
[26] Pliego, J. R. Riveros, J. M. *Phys. Chem. Chem. Phys.,* 2004, *4,* 1622
[27] Kolthoff, I. M., Chantooni, M. K. *J. Am. Chem. Soc.,* 1969, *91,* 4621
[28] Pliego, J. R., Riveros, J. M. *Phys. Chem. Chem. Phys.,* 2002, *4,* 1622
[29] Jones, J. R. 'The Ionisation of Carbon Acids', Academic Press, London, 1973
[30] Cox, R. A., Stewart, R. *J. Am. Chem. Soc.,* 1976, *98,* 488
[31] Stewart, R. 'The Proton: Applications to Organic Chemistry' Academic Press, 1985, Ch 2
[32] Bordwell, F. G. Algrim, D. J. *J. Am. Chem. Soc.,* 1988, *110,* 2964
[33] Bordwell, F. G., Guo-Zhen, J. *J. Am. Chem. Soc.,* 1991, *113,* 8398
[34] Bell, R. P. 'The Proton in Chemistry', Chapman and Hall, 1973

[35] Eigen, M. *Angew. Chem. Int. Edit. Engl.,* 1964, *3*, 1
[36] Kresge, A. J. *Acc. Chem. Res.,* 1990, *23*, 43
[37] Amyes, T. L., Richard, J. P. *J. Am. Chem. Soc.,* 1996, *118*, 3129
[38] Kolthoff, I. M., Chantooni, M. K., Bhowmik, S. *J. Am. Chem. Soc.,* 1968, *90*, 23
[39] Asghar, B. H. M., Crampton, M. R. *Org. Biomol. Chem.,* 2005, *3*, 3971
[40] Crampton, M. R., Robotham, I. A. *J. Chem. Research (S),* 1997, 22
[41] Mucci, A., Domain, R., Benoit, R. L. *Can. J. Chem.,* 1980, 58, 953
[42] Benoit, R. L., Mackinnon, M. J., Bergeron, L. *Can. J. Chem.,* 1981, 59, 1501
[43] Brauman, J. I., Riveros, J. M., Blair, L. K. *J. Am. Chem. Soc.,* 1971, 93, 3914
[44] Hughes, D. L., Bergan, J. J., Grabowski, E. J. J. *J. Org. Chem.,* 1986, 51, 2579
[45] Bordwell, F. G., Branca, J. C., Hughes, D. L. Olmstead, W. N. *J. Org. Chem.,* 1980, 45, 3305
[46] Maran, F., Celadon, D., Severin, M. G., Vianello, E. *J. Am. Chem. Soc.,* 1991, *113*, 9320
[47] Pytela, O.; Kulhánek, J. *Collect. Czech. Chem. Commun.,* 2002, *67*, 596
[48] Ludwig, M., Baron, V., Kalfus, K., Pytela, O., Vecera, M. *Collect. Czech. Chem. Commun.,* 1985, *51*, 2135
[49] Roletto, E., Juillard, J. *J. Soln. Chem.,* 1974, *3*, 127
[50] Breant, M., Georges, J. *J. Electroanal. Chem.,* 1976, *68*, 165
[51] Breant, M., Auroux, A., Lavergne, M. *Anal. Chim. Acta,* 1976, *83*, 49
[52] Kolthoff, I. M., Chantooni, M. K., Smagowski, H. *Anal. Chem.,* 1970, *42*, 1622
[53] Izutsu, K., Nakamura, T., Takizawa, K., Takeda, A. *Bull. Chem. Soc., Japan,* 1985, *58*, 455
[54] Roletto, E., Vanni, A. *Talanta,* 1977, *24*, 73
[55] Mermet-Dupin, M. *J. Electroanal. Chem.,* 1974, *52*, 75
[56] Ji, P., Atherton, J. H., Page, M. I. *J. Org. Chem.,* 2011, *76*, 3286
[57] Ji, P., Atherton, J. H., Page, M. I. *Faraday Discussions,* 2010, 145, 15
[58] Herlem, M., Popov, A. I. *J. Am. Chem. Soc.,* 1972, *94*, 1431
[59] Tongraar, A., Kerdcharoen, T., Hannongbua, S. *J. Phys. Chem. A,* 2006, *44*, 4924
[60] Plambeck, J. A. *Can. J. Chem.,* 1969, *47*, 1401
[61] Ji, P., Atherton, J. H., Page, M.I *J. Org. Chem.,* 2011, *76*, 1425
[62] Ji, P, Powles, N. T., Atherton, J. H., Page, M. I. *Org. Lett.,* 2011, *13*, 6118
[63] Perrin, D. D., Dempsey, B., Serjeant, E. P. 'pK_a Prediction for Organic Acids and Bases' Chapman and Hall, London, 1981
[64] Exner, O. 'Advances in Linear Free Energy Relationships', Chapman, N. B.; Shorter, J.(Eds)., Plenum: New York, 1972, Chapter 2

Low-Basicity and Low-Polarity Aprotic Solvents

7

According to Eigen [1] the kinetics of dissociation and formation of acids are best described in terms of eq. (7.1) (for acid HA), in which $H^+ \ldots A^-$ is an ion-pair intermediate.

$$HA \rightleftharpoons H^+ \ldots A^- \rightleftharpoons H^+ + A^- \quad (7.1)$$

Similarly, proton transfer between acid, HA, and base, B, involves both hydrogen-bonded and ion-paired intermediates, eq. (7.2).

$$HA + B \rightleftharpoons AH \cdots B \rightleftharpoons A^- \cdots HB^+ \rightleftharpoons A^- + BH^+ \quad (7.2)$$

Under normal circumstances, in aqueous solution, the stationary concentrations of the intermediate ion-pair species are invariably very small, and hence they make no effective contribution to the *thermodynamics* of ionization or proton-transfer equilibria. Similarly, in the more highly polar and basic aprotic solvents (Chapter 6) it is normally possible to find conditions under which ion-paired or hydrogen-bonded species are either at low concentration or can be allowed for in a relatively straightforward manner. At the other extreme of very low-polarity solvents, such as chloroform and carbon tetrachloride, however, reaction between acetic acid (HA) and triethylamine (B) has been shown by infrared spectroscopy to generate significant levels of ion-pairs, but no measurable concentrations of free ions [2, 3]. Proton-transfer equilibria in such non-dissociating solvents are discussed in the following chapter.

In this chapter we consider the dissociation of acids in a range of low-basicity and low-polarity aprotic solvents, such as acetonitrile (MeCN), 4-methyl-1,3-dioxolane (propylene carbonate, PC), nitromethane (NM), acetone, tetrahydrofuran (THF), tetrahydrothiophene 1,1-dioxide (sulpholane, TMS), methyl isobutyl ketone (MIBK), and nitrobenzene. They range in dielectric constant from $\varepsilon_r = 64.9$ (PC) to 7.6 (THF), but mostly have values in the region 20–40. The most prominent solvents in terms of the data available are MeCN and THF.

Solvents in this group, in common with all aprotic solvents, are poor at solvating anions, but they are also poor at solvating the proton and simple cations, because of their low basicity. This is apparent from an examination of the solvation of the ions (Section 3.3.2): large increases in free energies are observed for both simple cations and anions on transfer from water to all of these solvents. This leads to very low dissociation constants of all acids, and especially neutral acids, compared with those in water and the more basic aprotic solvents, such as dimethylsuphoxide (DMSO), dimethylformamide (DMF) and N-methylpyrrolidin-2-one (NMP).

There are additional consequences of the high activities of ions in these solvents. First is a strong tendency towards ion-pair formation and hydrogen-bond association, which means that acid-base equilibria at practical concentrations are frequently dominated by the various association equilibria; this is treated more quantitatively in Chapter 8. Secondly, experimental measurements are often very susceptible to the purity of the solvents, and especially to residual water levels. In acetonitrile, for example, equilibrium constants for the hydration of the proton, $K_h(H_2O)_n$, eq. (7.3) [4], for $n = 1$ to 4 are successively, 1.6×10^2 M^{-1}, 8×10^3 M^{-2}, 6×10^4 M^{-3}, and 2×10^5 M^{-4} [4].

$$H^+ + nH_2O \xrightleftharpoons{K_h(H_2O)_n} H(H_2O)_n^+ \qquad (7.3)$$

Even small amounts of water, therefore, are sufficient to reduce the proton activity substantially. Association between ions and water molecules has also been demonstrated in propylene carbonate by NMR studies [5].

Low levels of water can also significantly influence the *relative* strengths of acids, because of differing responses of anions to hydration, depending upon the extent of charge delocalization of the anions [6].

These factors can make the establishment of an absolute scale of acidities difficult, and the first step is often to obtain a reliable set of *relative* acidities, based on, for example, spectrophotometric measurement of equilibrium constants between acids of similar acidity, or potentiometric measurements related to a common standard solution (see below). In most cases, a more comprehensive set of data exists for cationic acids, such as protonated amines, which typically have higher dissociation constants and are less susceptible to homohydrogen-bond formation.

Au: as before, query with author.

7.1 Acetonitrile

Acetonitrile is both weakly basic and weakly acidic, with relatively low Donor and Acceptor Numbers (Section 1.2), but it has a sufficiently high dielectric constant (37.5) to limit the extent of ion-pair formation. The autoprotolysis constant is very low [7]: $pK_{IP} \geq 33^*$, so that extremely strong bases can be tolerated without deprotonation of the solvent.

There is a considerable body of data available in acetonitrile, which has recently been carefully revised and extended [8–19]. Of particular interest has been the introduction the phosphazene bases, which are strong, neutral bases, capable of ionizing a variety of carbon acids (Section 7.1.2).

*It is likely to be considerably higher than this, as CH$_3$CN has $pK_a = 31.3$ in DMSO, which would suggest a corresponding value in acetonitrile as solvent of significantly greater than 40 (Section 7.1.1)

7.1.1 Neutral acids: carboxylic acids and phenols, carbon acids

Leito, Koppel, and co-workers have recently reported a comprehensive, self-consistent acidity scale for neutral acids [19]. Their methodology is based on the spectrophotometric determination of the relative acidities of overlapping pairs of acids, HA_1 and HA_2, according to the equilibrium, eq. (7.4), thereby avoiding the need to determine $[H^+]$.

$$HA_2 + A_1^- \xrightleftharpoons{K_e} A_2^- + HA_1 \qquad (7.4)$$

The equilibrium constant, K_e, which is related to the two pK_a-values by eq. (7.5),[†] can be obtained from simple determination of any one of the components in eq. (7.4); the remaining values following from the stoichiometry of the system.

[†] Concentrations rather than activities can be used as the various activity coefficients cancel to a very good approximation

$$\log K_e = pK_{a1} - pK_{a2} = \log \frac{[A_2^-][HA_1]}{[HA_2][A_1^-]} \qquad (7.5)$$

The acids include carboxylic acids, phenols, alcohols, sulphonic acids, NH-acids, and carbon acids, and each was involved in at least two independent equilibria. Carbon acids were particularly useful in constructing the scale because they do not undergo homohydrogen-bond processes and their acidities are relatively independent of the presence of low levels of water. The spectrophotometric method typically also involves the use of very low concentrations, thus minimizing processes of ion association.

The resultant scale, covering 24 orders of magnitude, relates in the first instance of course only to relative acidities; the assignment of the absolute pK_a-values requires a reference compound, of known pK_a, which can be used to anchor the scale. Picric acid was used for this purpose. Its $pK_a = 11.1$ has been reliably determined by using three different methods—potentiometric, spectrophotometric, and conductimetric [9]—and it has the advantage of a very low homohydrogen-bond constant ($K_{AHA} = 2.4\,M^{-1}$) [20].

Agreement with earlier reported values for carboxylic acids and phenols [8–14], is mostly good, with the exception of the weakest acids, for which earlier studies appear to have generally slightly underestimated their pK_a-values. This is not unexpected, as the most basic anions are those which interact most strongly with any adventitious water, and is in keeping with the broad generalization that most of the error sources in non-aqueous measurements, lead to a contraction of the pK_a-scale. More generally, the susceptibility of the acid strength to the influence of trace water levels decreases in the order, OH-acids > NH-acids > CH-acids.

Carboxylic acids and phenols

Dissociation constants of carboxylic acids and phenols are listed in Table 7.1, and a more comprehensive list is included in Appendix 9.4.

Values from earlier studies [8–14] have been corrected according to recommendations by Leito and co-workers [19]; the corrections vary from zero for acids with $pK_a \sim 10$ to an increase of 1.2 units for phenol, $pK_a = 28.5$.

Fig. 7.1 shows a comparison of the dissociation constants with those in water.

The two series of acids form distinct correlations, represented by eq. (7.6) and (7.7); the larger slope for the phenols is indicative of the greater levelling of substituent effects in water, because of the strong dependence upon substituent

Table 7.1 pK_a-values of carboxylic acids and phenols in acetonitrile at 25°C[a]

Carboxylic acid	pK_a MeCN	$pK_a H_2O$	ΔpK_a[b]	Benzoic acid/Phenol	pK_a MeCN	$pK_a H_2O$	ΔpK_a[b]
dichloroacetic	16.4	1.34	15.1	4-chloro	20.9	4.00	16.9
2-chloroacetic	19.7	2.85	16.9	3-methoxy	21.3	4.12	17.2
acetic	23.1	4.75	18.3	H	21.5	4.19	17.3
				3-methyl	21.5	4.28	17.2
Benzoic acid:				4-methyl	21.9	4.38	17.5
2,6-dinitro	16.2	1.14	15.1	4-hydroxy	21.6	4.55	17.1
2,4-dinitro	16.6	1.43	15.2				
2,6-dichloro	18.2	2.66	15.5	Phenol:			
2-nitro	18.8	2.19	16.6	2,4,6-trinitro	11.0	0.43	10.6
3,5-dinitro	17.7	2.67	15.0	2,4-dinitro	16.4	4.10	12.3
3,4-dinitro	18.0	2.82	15.2	3,5-dinitro	21.3	6.66	14.6
2-chloro	19.7	2.92	16.8	2-nitro	23.0	7.23	15.7
4-nitro	19.3	3.43	15.9	4-nitro	21.7	7.23	14.5
3-nitro	20.2	3.47	16.7	3-nitro	24.9	8.36	16.5
4-cyano	19.9	3.53	16.4	3,4-dichloro	25.1	8.51	16.6
3-cyano	20.0	3.60	16.4	4-cyano	23.7	8.58	15.1
3-chloro	20.1	3.80	16.3	3-chloro	26.1	9.02	17.1
3-bromo	20.2	3.83	16.4	4-bromo	26.8	9.36	17.4
4-bromo	20.0	3.99	16.0	H	28.5	9.99	18.5

[a] Ref. [8–14, 19]; [b] $\Delta pK_a = pK_a(MeCN) - pK_a(H_2O)$

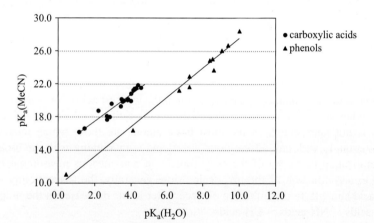

Fig. 7.1.
pK_a-values of carboxylic acids and phenols in acetonitrile versus water at 25°C

of the charge density of the phenoxide ions, and hence their solvation by water, as discussed previously (Section 6.1.1).

$$\text{carboxylic acids}: pK_a(MeCN) = 1.6 pK_a(H_2O) + 14.9 \tag{7.6}$$

$$\text{phenols}: pK_a(MeCN) = 1.8 pK_a(H_2O) + 9.6 \tag{7.7}$$

The absolute increases of both series on transfer from water are very large, e.g., > 18 log units for both acetic acid and phenol, but are entirely to be expected on the basis of the solvation of ions in acetonitrile relative to water (Table 3.7). Thus, for example, *the combined increase in free energy for*

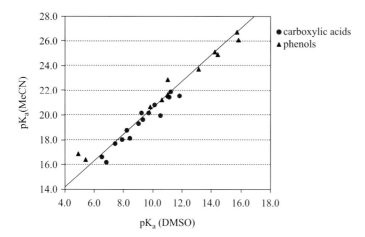

Fig. 7.2.
pK$_a$-values of carboxylic acids and phenols in acetonitrile versus dimethylsulfoxide at 25°C

H$^+$ and OAc^- on transfer from water to acetonitrile of 105.7 kJ mol^{-1} is equivalent to 18.5 pK-units.

When the values are compared with those measured in DMSO (Section 6.1.1), however, the distinction between the two series of acids (carboxylic acids and phenols) disappears, Fig. 7.2. This has previously been observed for the transfer of neutral acids from DMSO to various basic aprotic solvents, including DMF and NMP, and is attributable to the lack of specific solvation of anions in aprotic solvents.

The best-fit correlation line is given by eq. (7.8) and corresponds to a constant increase in pK$_a$ of 10.5 units, irrespective of the acid.

$$pK_a(MeCN) = 1.00 pK_a(DMSO) + 10.5 \qquad (7.8)$$

Among the phenols, positive deviations from the correlation are observed for the 2-nitro-, 2,6-dinitro-, and 2,4,6-trinitro-derivatives, and the reduced tendency of these substrates to dissociate compared with the other acids may be attributed to stabilization of their acid forms in acetonitrile by intramolecular H-bonding between the adjacent –OH and –NO$_2$ groups.

In terms of the solvation of the ions alone, an increase in pK$_a$ from DMSO to MeCN of closer to 11.5 units would be expected, for example, $\Delta G_{tr}(H^+ + OAc^-) = 64$ kJ mol^{-1}, 11.3 log units, but the effect is mitigated by the stronger solvation of the acid molecules by the more basic DMSO; this typically contributes up to 10 kJ mol^{-1} to the stability of the acid (Section 3.3.2 and 3.4).

The lack of H-bonding between MeCN and carboxylic acids and phenols is also apparent in two other phenomena: strong homohydrogen-bond equilibria, eq. (7.9), for which $K_{AHA} \sim 10^4$ M^{-1} and $\sim 5 \times 10^3$ M^{-1} for phenols and carboxylic acids, respectively; and very effective intramolecular hydrogen-bonding of the mono-anion of suitable dicarboxylic acids, notably malonic acid, eq. (7.10), for which $K' = 2.5 \times 10^4$ [21].

$$HA + A^- \xrightleftharpoons{K_{AHA}} A^- \cdots HA \tag{7.9}$$

(7.10)

Carbon acids

Carbon acids, in contrast to the carboxylic acids and phenols, are unable to participate in H-bonding with either DMSO or MeCN, and thus are expected to show larger increases in pK_a compared with DMSO, and indeed their pK_a-values are typically around 12.9 units higher than those in DMSO (see Table 7.2). Measurements have been mostly confined to highly activated systems, because of the very weak acidity of simple carbon acids, such as ketones. Substrates reported include conjugated hydrocarbons, such as the fluorenes, and substituted alkanes containing a combination of nitrile and aromatic substituents [19].

It is most useful to compare the results for carbon acids with corresponding acidities in DMSO, for which an extensive set of data exists [22, 23]. There is a good correlation between the pK_a-values of the carbon acids in the two solvents, with those in acetonitrile being on average 12.9 units higher than those in DMSO [19]. The difference resides almost entirely in the decreased solvation of the proton in MeCN, estimated to contribute some 11.3 log units to the difference in pK_a (Table 3.7). Representative dissociation constants are listed in Table 7.2.

On the basis of the essentially constant difference between the dissociation constants in DMSO and MeCN, it is possible to estimate the pK_a-values of other important carbon acids, such as ketones and nitroalkanes, with some degree of confidence, and the estimated values have been included in the Table. It is apparent that ketones, in particular, will be very weak indeed, with

Table 7.2 pK_a-values of carbon acids in acetonitrile at 25°C[a]

Acid	pK_a(MeCN)	pK_a(DMSO)[b]	ΔpK_a[c]
$(C_6H_5)(C_6F_5)$CHCN	26.1	12.8	13.3
9–CO_2Me-fluorene	23.5	10.3	13.2
9-CN-fluorene	21.3	8.3	13.0
$(C_6F_5)_2$ CHCN	21.1	8.0	13.1
$(4-Me-C_6H_4)(C_6F_5)$CHCN	18.1	4.9	13.2
$4-Me-C_6H_4CH(CN)_2$	17.6	4.9	12.7
$4-Cl-C_6H_4CH(CN)_2$	17.4	4.5	12.9
$C_6F_5CH(CN)_2$	13.0	0.3	12.7
CH_3COCH_3	(39.4)[d]	26.5	
$C_6H_5COCH_3$	(37.6)[d]	24.7	
CH_3NO_2	(30.3)[d]	17.4	

[a] Ref. [19]; [b] Ref. [22, 23]; [c] $\Delta pK_a = pK_a$(MeCN)–pK_a(DMSO); [d] Estimated by adding 12.9 to DMSO values

pK_a > 35 for benzophenones and close to 40 for aliphatic ketones; thus even the strongest neutral bases will only be able to ionize them to a very limited extent (Chapter 8).

7.1.2 Cationic acids (neutral bases)

The dissociation constants of the conjugate acids of a large number of neutral bases in acetonitrile have been reported [24–26]. Experimental methods used include both potentiometric and spectrophotometric measurements, and equilibrium constants for an extensive series of overlapping pairs of bases have been measured spectrophotometrically by Leito and co-workers [18], in a manner analogous to that described above for the neutral acids.

The basicity scale covers some 28 units, with the strongest bases being the phosphazene bases, examples of which are given in Scheme 7.1. The defining feature of these bases is the $R_1N = P(R)_3$ unit, the number of multiples of which, for a given amine, determines the overall basicity [16, 17]. An extended list of phosphazene bases is included in Appendix 9.4.3.

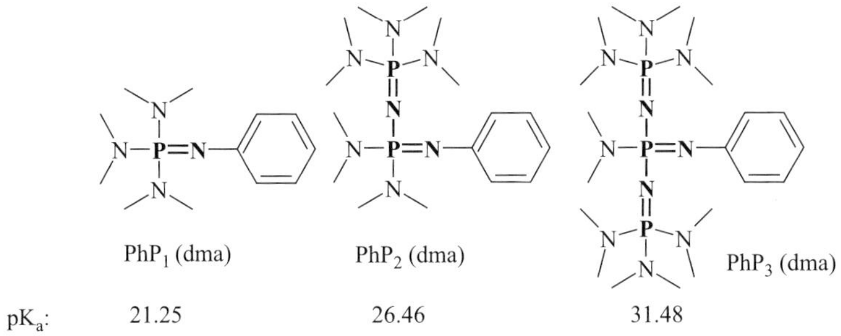

Scheme 7.1.
Phosphazene bases in acetonitrile [16–18, 18]

	PhP$_1$ (dma)	PhP$_2$ (dma)	PhP$_3$ (dma)
pK_a:	21.25	26.46	31.48

Among more traditional nitrogen bases, pyridines, anilines, and amines have been well studied, and the results are reported below. Homohydrogen-bond formation constants, K_{BHB}, for association between the amines and their conjugated acids (B $\cdots$ BH$^+$), are significantly lower than those for the carboxylic acids and phenols, being typically around 0–30 M^{-1}. Homohydrogen-bond formation increases as the strength of the base, B, and the number of hydrogen atoms in BH$^+$ increases; it is most extensive for lower aliphatic primary amines, but it is insignificant for non-cyclic tertiary amines, with the exception of Me$_3$N, for which $K_{BHB} = 6$ M^{-1} [24].

Pyridines

Dissociation constants of protonated pyridines in acetonitrile are listed in Table 7.3, along with the corresponding values in water.

The dissociation constants of pyridines are very sensitive to substituent in both water and acetonitrile, and there is an excellent correlation between the two sets of data, Fig. 7.3, with the best-fit line being given by eq. (7.11). The observed slope (1.34) shows that the dissociation constants in acetonitrile are somewhat more sensitive to substituents than in water.

Table 7.3 pK_a-values of pyridinium ions in acetonitrile at 25°C[a]

Pyridine	pK_a MeCN	pK_a H$_2$O	ΔpK_a^b	Pyridine	pK_a MeCN	pK_a H$_2$O	ΔpK_a^b
2-chloro	6.8	0.49	6.3	3-hydroxy	12.6	4.75	7.8
2-bromo	7.0	0.71	6.3	H	12.6	5.17	7.4
2-hydroxy	8.3	1.25	7.1	3-methyl	13.7	5.58	8.1
3-cyano	8.0	1.38	6.6	2-methyl	13.9	5.91	8.0
4-cyano	8.5	1.9	6.6	4-methyl	14.5	5.93	8.6
2-acetylo	9.6	2.84	6.8	3-amino	14.4	6.03	8.4
3-bromo	9.5	2.84	6.7	2-amino	14.7	6.66	8.0
3-chloro	10.0	2.84	7.2	4-amino	18.4	9.06	9.3
3-acetylo	10.8	3.55	7.3				

[a] Ref. [18, 26]; [b] $\Delta pK_a = pK_a(MeCN) - pK_a(H_2O)$

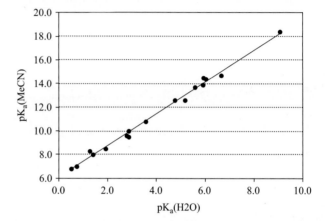

Fig. 7.3.
pK_a-values for pyridinium ions in acetonitrile versus water at 25°C

$$pK_a(MeCN) = 1.34 pK_a(H_2O) + 6.1 \quad (7.11)$$

Anilines

Dissociation constants of protonated anilines in acetonitrile are listed in Table 7.4, along with the corresponding values in water.

Table 7.4 pK_a-values of anilinium ions in acetonitrile at 25°C[a]

Aniline	pK_a MeCN	pK_a H$_2$O	ΔpK_a^b	Aniline	pK_a MeCN	pK_a H$_2$O	ΔpK_a^b
2-NO$_2$	4.80	-0.29	5.1	4-CF$_3$	8.03	2.75	5.3
2,6-Cl$_2$	6.06	0.42	5.6	2,4-F$_2$	8.39	3.26	5.1
2,5-Cl$_2$	6.21	1.61	4.6	4-Br	9.43	3.86	5.5
4-NO$_2$	6.22	0.99	5.2	2-Me	10.50	4.45	6.0
4-F-3-NO$_2$	7.67	2.36	5.3	H	10.82	4.60	6.2
3-NO$_2$	7.68	2.50	5.2	4-OMe	11.86	5.36	6.5
2-Cl	7.86	2.64	5.2				

[a] Ref. [18]; [b] $\Delta pK_a = pK_a(MeCN) - pK_a(H_2O)$

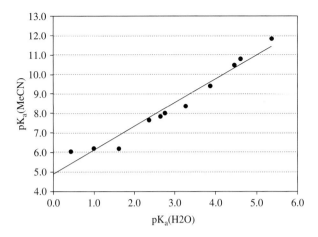

Fig. 7.4.
pK$_a$-values for anilinium ions in acetonitrile versus water at 25°C

There is again a good correlation between the two sets of data, shown in Fig. 7.4, with the best-fit line being given by eq. (7.12). The observed slope (1.22) again shows a greater sensitivity to substituent in MeCN than in water, but the effect is not as pronounced as for the pyridines.

$$pK_a(MeCN) = 1.22 pK_a(H_2O) + 4.9 \qquad (7.12)$$

Amines

Dissociation constants of protonated amines in acetonitrile are listed in Table 7.5, along with the corresponding values in water.

In the more basic aprotic solvents, there is a distinction in the behaviour of primary, secondary and tertiary amines on transfer from water, but any such effects are clearly absent in the results in Table 7.5; there is an almost constant increase in pK$_a$ of close to 7.7 units on transfer from water to acetonitrile.

The dominant factor determining the higher pK$_a$-values in MeCN of these cationic acids is the increase in the free energy of the proton. This may be shown from an analysis of the influence of MeCN on the dissociation constants

Table 7.5 pK$_a$-values of ammonium ions in acetonitrile at 25°C[a]

Amine	pK$_a$ MeCN	pK$_a$ H$_2$O	ΔpK$_a$[b]	Amine	pK$_a$ MeCN	pK$_a$ H$_2$O	ΔpK$_a$[b]
ammonia	16.5	9.21	7.3	pyrrolidine	19.6	11.27	8.3
methylamine	18.4	10.62	7.8	dimethylamine	18.7	10.6	8.1
ethylamine	18.4	10.63	7.8	diethylamine	18.8	10.98	7.8
n-propylamine	18.2	10.53	7.7	di-n-butylamine	18.3	11.25	7.1
n-butylamine	18.3	10.59	7.7	trimethlyamine	17.6	9.76	7.9
t-butylamine	18.1	10.45	7.7	triethylamine	18.5	10.76	7.8
benzylamine	16.9	9.34	7.6	tri-n-proplyamine	18.1	10.65	7.5
morpholine	16.6	8.36	8.2	tri-n-butylamine	18.1	10.89	7.2
piperidine	18.9	11.22	7.7				

[a] Ref. [24]; [b] ΔpK$_a$ = pK$_a$(MeCN)–pK$_a$(H$_2$O)

via eq. (7.13), in which $\Delta G_{tr}(H^+)$ represents the increase in free energy of the proton on transfer from water to acetonitrile, etc. (Section 3.7).

$$\Delta pK_a = pK_a(MeCN) - pK_a(H_2O)$$
$$= \{\Delta G_{tr}(H^+) + \Delta G_{tr}(B) - \Delta G_{tr}(BH^+)\}/2.303RT \quad (7.13)$$

The estimated value of $\Delta G_{tr}(H^+) = 45 \text{ kJ mol}^{-1}$ for transfer from water to acetonitrile is equivalent to an increase of 7.6 pK-units (Table 3.7). This is close to a broad average increase of ~ 7 units in pK_a across the three types of bases in Tables 7.3–7.5, which corresponds to an increase in the free energy of dissociation of 40 kJ mol^{-1}.

There are, however, significant variations amongst the different classes of cationic acids, which reflect corresponding changes in the relative free energies of transfer of the free and protonated base, $\Delta G_{tr}(B) - \Delta G_{tr}(BH^+)$. This term makes only a small contribution to the pK_a-values of the *aliphatic* amines, typically less than $\sim 5 \text{ kJ mol}^{-1}$ (1 pK-unit), but for the pyridinium and the anilinium ions there is a discernable trend with the pK_a of the acid; the stronger acids show smaller increases in pK_a-values. *The net result is that the spread in acidities across the whole range of protonated bases is higher in acetonitrile than in water.* The simplest explanation for this is that the acidities in water are attenuated by stabilization of the protonated bases by $NH^+ \cdots OH_2$ hydrogen-bonds; the effect is strongest for the most acidic protonated bases, thus reducing their relative acidity.

We may also compare the dissociation constants in MeCN with those in the more basic aprotic solvent DMSO. In all cases the pK_a-values are substantially higher in MeCN, with the largest increases being observed for the acids containing the more highly substituted nitrogen atoms. Thus, increases of 9.2 and 9.7 units for pyridinium and triethylammonium ions, respectively (which correspond closely to the expected increases in the free energy of the proton) may be compared with 7.7 and 7.1 units for the ethylammonium and anilinium ions, respectively. We have discussed earlier the increasingly strong interactions between the ammonium cations and DMSO as the number of N-H protons increases (Section 6.1.2), and these results are in accord with the loss of these specific interactions on transfer to acetonitrile. The dominant contribution to the differences in pK_a-values between DMSO and MeCN though is, of course, the increased activity of the proton in MeCN relative to DMSO.

7.2 Propylene carbonate, sulpholane, acetone, methyl *iso*-butyl ketone, nitrobenzene

This group comprises solvents which, like acetonitrile, mostly have sufficiently high dielectric constants to enable free ions to predominate up to reasonable concentrations (≤ 0.01 M). Amongst these, methyl *iso*-butyl ketone (MIBK) has the lowest dielectric constant ($\varepsilon_r = 12.9$) and ion-pair formation is extensive, even at low concentrations;* for example, anilinium perchlorate has an ion-pair formation constant, $K_{IP} = 3.8 \times 10^4 \text{ M}^{-1}$ [27]. By contrast, ion-pair

*The dependence upon K_{IP} of the extent of ion-pair formation for different concentrations of salt, MX, is illustrated in Table 4.2

formation constants in nitrobenzene ($\varepsilon_r = 34.8$) are typically $< 50\,\mathrm{M}^{-1}$ [28]. Acetone is expected to be the most effective at solvating the proton, because of favourable Donor Number, β-value, and observed free energies of ion solvation (Sections 1.2 and Section 3.3.2), and MIBK and nitrobenzene the poorest at solvating ions in general. Propylene carbonate has a high dielectric constant ($\varepsilon_r = 64.9$), but other solvent and ion-solvating properties are similar to those of MeCN.

In contrast to the detailed and very reliable results available in aprotic solvents such as DMSO and MeCN, there are often significant uncertainties in results in this group of solvents [29]; they are mostly associated with problems arising in the standardization of the glass electrodes used in potentiometric measurements. The use and applications of potentiometric ion sensors in non-aqueous solvents has been reviewed by Coetzee and co-workers [30, 31].

A selection of dissociation constants is listed in Table 7.6, which includes also the corresponding values in water and MeCN for comparison. The data is limited in scope, but is generally sufficient to be indicative of trends to be expected for these solvents when compared with water and MeCN (and other aprotic solvents).

Acid strengths of neutral acids in PC are similar to, but slightly higher than in MeCN; the same is true for the *n*-butylammonium ion ($pK_a(PC) = 17.0$, $pK_a(MeCN) = 18.3$) [32]. pK_a-values for a series of substituted benzoic acids in acetone are consistently several units lower than those in MeCN, and the same is true for protonated *n*-butylamine, which is some 6 pK-units more acidic in acetone [33]. The results are consistent with acetone being more basic than MeCN and also MIBK [34].

The most surprising results are those for nitrobenzene as solvent [35], for which the reported acidities are considerably higher than those of the other solvents, despite the fact that the low Donor and Acceptor Numbers and the unfavourable thermodynamics of solution of simple electrolytes in nitrobenzene [28] would suggest very poor ion-solvation compared with the other aprotic solvents in Table 7.6. The pK_a-measurements are anchored to a careful spectrophotometric measurement of the dissociation constant of picric acid, but some doubt must remain as to the reliability of the absolute values.

Table 7.6 pK_a-values in propylene carbonate (PC), acetone, methyl-*iso*-butyl ketone (MIBK), and nitrobenzene ($PhNO_2$) at 25°C

Acid	pK_aH_2O	pK_aMeCN	pK_aPC^a	pK_a acetone[b]	pK_aMIBK^c	$pK_aPhNO_2^d$
perchloric	strong	2.0	1.3		4.5	
methanesulfonic	−1.6	10.0	8.3			
picric	0.43	11.0	9.3	9.2	11.0	6.6
benzoic	4.19	21.5	19.7	18.2		
anilinium	4.63	10.6		5.9	9.6	5.2
pyridinium	5.25	12.6		7.2		7.5

[a] Ref. [32]; [b] Ref. [33]; [c] Ref. [34]; [d] Ref. [35]

A wide range of substituted benzoic acids have been investigated in sulfolane [36], with reported dissociation constants being typically 6 pK-units higher than in acetonitrile. There is again a question associated with the absolute values, however, as they are referred to a value for picric acid, $pK_a = 17.4$ [37], about which there is considerable uncertainty [30, 38].

Extensive compilations of dissociation constants of substituted pyridinium ions, eq. (7.14), in several solvents exist [26], summarized in Table 7.7 and Fig. 7.5. Table 7.7 includes values for water and acetonitrile for comparison.

$$PyH^+ \xrightleftharpoons{K_a} H^+ + Py \qquad (7.14)$$

The best-fit lines in Fig. 7.5 can be represented by eq. (7.15).

$$pK_a(S) = m pK_a(H_2O) + c \qquad (7.15)$$

Table 7.7 pK_a-values of pyridinium ions in nitromethane (MeNO$_2$), propylene carbonate (PC), and acetone at 25°C[a]

Pyridine	pK_aH$_2$O	pK_a MeCN	pK_aMeNO$_2$	pK_a PC	pK_a acetone
2-Cl	0.48	6.80	5.8	5.51	2.96
2-Br	0.66	7.02	6.61		3.0
2-OH	1.22	8.26	7.22		4.60
3-CN	1.4	8.04	7.2	7.12	3.7
4-CN	1.9	8.50	7.66	7.34	4.37
3-Cl	2.84	10.01	9.1	8.28	4.42
3-Br	2.84	9.49	9.60		4.2
3-OAc	3.26	10.75	10.1		4.93
3-OH	4.75	12.63	11.6		7.55
H	5.17	12.60	12.23	11.54	7.23
3-Me	5.58	13.66	13.23	11.16	7.59
4-NH$_2$	9.06	18.38	17.67	16.42	12.69

[a] Ref. [26]

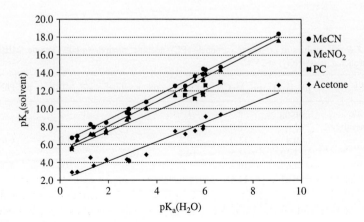

Fig. 7.5.
pK_a-values for pyridinium ions in solvents versus water at 25°C

The slopes, m, and intercepts, c, for the various solvents are:

Solvent	m	c
Acetonitrile	1.34	6.10
Nitromethane	1.37	5.33
Propylene carbonate	1.15	5.19
Acetone	1.08	1.99

The differences between the solvents should be dominated by their relative basicity with respect to the proton, which both in terms of proton solvation and solvent Donor Number should be strongest for acetone; the observed lower pK_a-values in acetone, Table 7.7, Fig. 7.5 are entirely consistent with this. The free energies of Py and PyH$^+$, the other components in the dissociation, eq. (7.14), will be a lot less susceptible to solvent variation than the proton.

7.3 Tetrahydrofuran

Tetrahydrofuran (THF) may be regarded as the prototype for a class of saturated cyclic ethers—a group of solvents that has exceptional importance in chemical research and in industrial processes [39]. It is used extensively as a solvent for polyvinyl chloride in printing inks, lacquers, and adhesives, and both THF and its 2-methyl derivative are important solvents for lithium batteries. In addition, it is a very useful solvent for reactions of organoalkali compounds (metalation reactions) in synthetic chemistry.* Its aprotic nature means that it is relatively inert to alkali metals and other strong reducing agents and bases. Furthermore, despite its low polarity ($\varepsilon_r = 7.6$), it is an excellent solvent for salts such as LiAsF$_6$, largely because of its relatively strong solvation of lithium ions, which leads to a high solubility and reasonably high electrical conductivity. Importantly, the response of the glass electrode in THF is Nernstian over a wide range of proton activities.†

The major difficulty associated with establishing a quantitative scale of acid–base chemistry in THF is very high ion-pair formation constants, K_{IP}, which are commonly $> 10^6$ M^{-1}. Thus, for example, for tetrabutylammonium benzoate, eq. (7.16), $K_{IP} = 7.4 \times 10^6$ M^{-1} [40].

$$Bu_4N^+ + PhCO_2^- \underset{}{\overset{K_{IP}}{\rightleftharpoons}} (Bu_4N^+PhCO_2^-) \quad (7.16)$$

Furthermore, at concentrations of around 10^{-3}M and higher, terniary ion formation ($[(Bu_4N^+)_2PhCO_2^-]$ and $[Bu_4N^+(PhCO_2^-)_2]$) must also be considered [41]. This means that absolute determinations of dissociation constants have to be made at very low concentrations ($\sim 10^{-5}$M) and even then need to be combined with corrections for the effect of ion-pairing on the acid-base ratios (Appendix 4.2).

One approach to this problem, favoured especially by Streitwieser and co-workers [42–44], is to recognize that at any practical concentrations ion-pairs will dominate and hence to derive a set of 'ion-pair' acidities. This has been

*Detailed structural and mechanistic studies of organolithium reagents in organic chemistry have been reported by Collum and co-workers: Collum, D.B.; McNeil, A.J.; Ramirez, A. *Angew. Chem., Int. Ed.* 2007, 49, 3002; Lucht, B.L.; Collum, D.B. *Acc. Chem. Res.*, 1999, 32, 1035

†Nernstian behaviour corresponds to a change in potential of $RT/(2.303F) = 0.059$ V/unit change in pH at 25°C. This follows from the Nernst Equation for the hydrogen electrode, $E = E^o + (RT/F)\ln[H^+]$ at $p(H_2) = 1$ atm, i.e., $E = E^o - 0.059$pH

done for both caesium ion-pairs, which involve direct contact between the cation and anion (contact ion-pairs), and lithium ion-pairs, which predominantly involve solvent-separated ion-pairs, in which the lithium ion retains a full coordination sphere of THF molecules and hence interacts more weakly with the anion. Eq. (7.17) shows the resulting proton-transfer equilibrium between two acids, RH and R'H, for cesium ion-pairs and similarly for lithium ion-pairs.

$$RH + Cs^+R'^- \xrightleftharpoons{K} Cs^+R^- + R'H \qquad (7.17)$$

The equilibrium constants, measured spectrophotometrically, were then converted to a numerical pK-scale, via eq. (7.18), by arbitrarily assigning to the caesium (or lithium) ion-pair of fluorene its free-ion pK_a-value in DMSO of 22.90.

$$pK_a(RH) = pK_a(R'H) - \log K \qquad (7.18)$$

R'H = fluorene

There is an excellent correlation between the lithium and caesium scales, and the observed slope (1.08) shows that the caesium scale is slightly more compressed than the lithium scale. This result is probably due to slightly higher electrostatic attraction of the carbanion to the caesium cation, which is effectively smaller than the solvated lithium ion [43]. There is also an excellent correlation with the absolute pK_a-values of the various carbon acids in DMSO, with a slope close to unity.

Both lithium and caesium enolates are known to aggregate in THF [45, 46] and, for example, the lithium enolate of p-phenylisobutyrophenone (LiSIBP) has a dimerization constant of 5×10^4 M^{-1} [45].

LiSIBP

The alkylation of p-t-butylbenzyl bromide by the lithium enolate, monomer/dimer mixture appears to proceed via the monomer, despite its low equilibrium proportion at higher concentrations.

The lithium and caesium enolates of dibenzyl ketone both exist as contact ion-pairs. These have a strong tendency to dimerize, as shown by UV/Vis spectral measurements and an apparent increase in acidity at higher concentrations [47]. The corresponding dianions can also be formed, and the first and second caesium pK_a-values are 18.07 and 33.70, respectively; very slow deprotonation kinetics were observed when using lithium bases, and the second pK_a-values were not able to be measured.

It is possible in a similar manner to derive a set of ion-pair basicities of amines in THF, eq. (7.19), in which B is a base, such as an amine, and HIn is a suitable proton donor, generally an indicator of known relative acidity [48].

$$B + HIn \underset{}{\overset{K_a(IP)}{\rightleftharpoons}} BH^+In^- \underset{}{\overset{(K_{IP})^{-1}}{\rightleftharpoons}} BH^+ + In^- \quad (7.19)$$

Unlike the caesium or lithium ion-pair acidities of neutral acids, however, the ion-pair basicities of the amines show only a rough correlation with ionic pK_a-values for BH^+ in MeCN and DMSO, and they cannot be placed on any of the ion-pair acidity scales for neutral acids in THF.

Conventional dissociation constants for a range of benzoic acids, phenols and protonated amines in THF have been reported more recently [35, 40, 49–52]. Picric acid ($pK_a = 11.84$) has been recommended as the preferred standard for calibration of the glass electrode, used in conjunction with a reference electrode of saturated $AgNO_3/Ag$ in THF [41]. In a typical procedure for the determination of pK_a-values of neutral acids, the acids in THF were titrated with aqueous Bu_4NOH and the pH-values recorded. Conductimetric titrations were then used to determine the ion-pair formation constants (normally $> 10^5 \, M^{-1}$, and up to $10^7 \, M^{-1}$), a knowledge of which enabled calculation of the concentrations of free ions. The increasing levels of water added with the titrant meant that the calculated pK_a-values changed systematically during the titrations, but the values in pure THF were determined by linear extrapolation to zero water content of the solvent. The choice of picric acid as a reference standard was based on the fact that amongst various possible standards for electrode calibration, it is the least sensitive to changes in solvent composition.

Similar methods were used for the titration of nitrogen bases, using 70% aqueous perchloric acid as the titrant [50]. K_{IP}-values (eq. (7.16)) for anilinium perchlorates vary in the range $4 \times 10^5 \, M^{-1} < K_{IP} < 5 \times 10^6 \, M^{-1}$, and formation constants for triple-ions from the ion-pairs are typically $\sim 10^4 \, M^{-1}$. Relative basicities for a wide set of phosphazines and N-bases (pyridines, amines, amidines) were determined using the procedures developed for measurements in MeCN [50], and absolute pK_a-values were tentatively anchored to an estimated pK_a-value for triethylamine ($pK_a = 12.5$). More recently an absolute pK_a-scale has been firmly established through a combination of potentiometric and conductimetric measurements at low concentrations on eleven bases (amines, anilines, pyrrolidines and iminophosphoranes), and this has allowed the calculation of absolute pK_a-values for 77 bases from the earlier relative pK_a-data [51]. A relative acidity scale for phosphorous-containing compounds has also been determined from measurements of a wide range of equilibrium constants using ^1H and ^{31}P NMR [53].

Dissociation constants for benzoic acids and phenols are reported in Table 7.8, along with the corresponding values in water for comparison, and those for the conjugate acids of selected neutral bases in Table 7.9. A more comprehensive listing is included in Appendix 9.5.

Although there is little thermodynamic data available on the solvation of ions in THF, we can get an indication of the behaviour to be expected from the acid and base properties of the solvent in comparison to MeCN and DMSO, Section 1.2, reproduced here for convenience:

Solvent	Donor Number DN[a]	H-bond basicity β^a	Acceptor Number AN[a]
DMSO	29.8	0.76	19.3
THF	20.0	0.55	8.0
MeCN	14.1	0.40	18.9

[a] See Section 1.2

Table 7.8. pK_a-values of carboxylic acids and phenols in tetrahydrofuran at 25°C[a]

Benzoic acid	pK_a THF	$pK_a H_2O$	ΔpK_a^b	Phenol	pK_a THF	$pK_a H_2O$	ΔpK_a^b
2-nitro	21.10	2.21	18.9	2,4,6-trinitro	11.84	0.30	11.5
3,5-dinitro	18.99	2.82	16.2	4-nitro	21.13	7.20	13.9
4-nitro	21.16	3.44	17.7	2-nitro	24.41	7.20	17.2
3-nitro	21.77	3.45	18.3	3,5-dichloro	23.16	8.30	14.9
3,5-dichloro	21.64	3.50	18.1	3-nitro	23.76	8.30	15.5
3-bromo	23.23	3.81	19.3	2-chloro	26.30	8.38	17.9
4-chloro	23.88	3.99	19.9	4-bromo	27.30	9.34	18.0
H	25.11	4.21	20.9	4-chloro	26.80	9.42	17.4
3-methyl	25.34	4.27	21.1	H	29.23	9.99	19.2

[a] Ref. [40, 49]; [b] $\Delta pK_a = pK_a(THF) - pK_a(H_2O)$

Table 7.9. pK_a-values of conjugate acids of neutral bases in tetrahydrofuran at 25°C[a]

Base	pK_a THF	$pK_a H_2O$	ΔpK_a^b	Base	pK_a THF	$pK_a H_2O$	ΔpK_a^b
$PhP_3(pyrr)^c$	26.8			pyridine	8.25	5.17	3.1
$PhP_3(dma)^c$	26.2						
$PhP_2(dma)^c$	22.2			Aniline:			
$PhP_1(dma)^c$	17.8			H	7.97	4.60	3.4
pyrrolidone	15.6	11.27	4.3	4-Cl	6.97	3.98	3.0
triethylamine	13.7	10.67	3.0	3-Cl	6.38	3.51	2.9
propylamine	14.7	10.70	4.0	3-NO_2	5.81	2.50	3.3
DMAP[d]	14.1	9.60	4.5	4-NO_2	4.82	0.99	3.8

[a] Ref. [51, 52]; [b] $\Delta pK_a = pK_a(THF) - pK_a(H_2O)$; [c] Phosphazine base, Scheme 7.1, pyrr = pyrrole; [d] 4-dimethylamino pyridine

It is apparent that in terms of solvent basicity, as measured by either by the Donor Number, DN, or the H-bond basicity, β, THF occupies an intermediate position between that of MeCN (low) and DMSO (high). On this basis the dissociation constants of the *cationic* acids in THF (Table 7.8) are expected to be intermediate between those in DMSO and MeCN, because the solvation of the proton largely dominates the solvent dependence of the acidity of cationic acids. In practice this is indeed the case, taking for example the pK_a-values of the triethylammonium ion in the three solvents: $pK_a(DMSO) = 9.0$ (Table 6.6) $< pK_a(THF) = 14.9$ (Table 7.9) $< pK_a(MeCN) = 18.5$ (Table 7.5). Compared with water, nitrogen bases in THF are on average ∼ 3.5 units more basic.

A striking feature of THF is the very low Acceptor Number compared with the majority of aprotic solvents that we have considered. This would suggest very poor anion solvation, with correspondingly high pK_a-values for neutral acids and a greater sensitivity to substituents than is found for other aprotic solvents. Both effects are observed in practice, more strongly so for benzoic acids than for phenols, no doubt because of the more dispersed charge on the phenoxide anions. Bosch, Barrón, and co-workers [40, 49] have discussed substituent effects for aromatic carboxylic acids and phenols in THF in some detail, and two features are notable, both being attributable to the poor anion solvation in THF. The first is a stronger sensitivity to substituent than in both DMSO and MeCN and (especially) also in water. The second is a much greater deviation of 2-substituted acids from correlations based on substituents in positions 3–5.

The large increases in pK_a-values relative to water are immediately apparent from Table 7.8, but it is instructive also to compare values in THF with those in DMSO and MeCN:

Acid	pK_aH$_2$O	pK_aDMSO[a]	pK_aMeCN[b]	pK_aTHF[c]
benzoic	4.19	11.1	21.5	25.1
phenol	9.99	18.0	28.5	29.2

[a] Tables 6.1, 6.3; [b] Table 7.1; [c] Table 7.8

DMSO is a better solvent for both anions and cations than is THF, and this is reflected in much lower pK_a-values for both benzoic acid and phenol. Compared with MeCN, THF is more basic and a better solvent for the proton, as discussed above, but despite this the pK_a-values are higher in THF because of overriding poor anion solvation.

References

[1] Eigen, M. *Angew. Chem. Int. Edit. Engl.,* 1964, *3*, 1
[2] Yerger, E. A., Barrow, G. M. *J. Am. Chem. Soc.,* 1955, *77*, 4474
[3] Barrow, G. M., Yerger, E. A. *J. Am. Chem. Soc.,* 1954, *76*, 5211
[4] Chantooni, M. K., Kolthoff, I. M. *J. Am. Chem. Soc.,* 1970, *92*, 2236
[5] Butler, J. N., Cogley, D. R., Grunwald, E. *J. Phys. Chem.,* 1971, *75*, 1477
[6] Kaupmees, K., Kaljurand, I., Leito, I. *J. Phys. Chem. A,* 2010, *114*, 11788
[7] Kolthoff, I. M., Chantooni, M. K. *J. Phys. Chem.,* 1968, *72*, 2270
[8] Coetzee, J. F., Padmanabhan, G. R. *J. Phys. Chem.,* 1965, *69*, 3193
[9] Kolthoff, I. M., Chantooni, M. K. *J. Am. Chem. Soc.,* 1965, *87*, 4428
[10] Kolthoff, I. M., Chantooni, M. K., Bhowmik, S. *J. Amer. Chem. Soc.,* 1966, *88*, 5430
[11] Kolthoff, I. M., Chantooni, M. K. *J. Phys. Chem.,* 1966, *70*, 856
[12] Kolthoff, I. M., Chantooni, M. K. *J. Am. Chem. Soc.,* 1969, *91*, 4621
[13] Chantooni, M. K., Kolthoff, I. M. *J. Phys. Chem.,* 1975, *79*, 1176
[14] Chantooni, M. K., Kolthoff, I. M. *J. Phys. Chem.,* 1976, *80*, 1306
[15] Pawlak, Z., Urbaczyk, G. *J. Mol. Struct.* 1988, *177*, 401
[16] Schwesinger, R., Willaredt, J., Schlemper, H., Schmitt, D., Fritz, H. *Chem. Ber.,* 1994, *127*, 2435
[17] Kaljurand, I., Rodima, T., Leito, I., Koppel, I. A., Schwesinger, R. *J. Org. Chem.,* 2000, *65*, 6202

[18] Kaljurand, I., Kütt, A., Sooväli, L., Rodimer, T., Mäemets, V., Leito, I., Koppel, I. A. *J. Org. Chem.*, 2005, *70*, 1019
[19] Kütt, A., Leito, I., Kaljurand, I., Sooväli. L., Vlasov, V. M., Yagupolskii, L. M., Koppel, I. A. *J. Org. Chem.*, 2006, *71*, 2829
[20] Kolthoff, I. M., Chantooni, M. K. *J. Phys. Chem.*, 1969, *73*, 4029
[21] Chantooni, M. K., Kolthoff, I. M. *J. Phys. Chem.*, 1975, *79*, 1176
[22] Bordwell, F. G. *Acc. Chem. Res.*, 1988, *21*, 456, and references therein
[23] Koppel, I. A., Koppel, J., Pihl, V., Leito, I., Mishima, M., Vlasov, V. M., Yagupolskii, L. M., Taft, R. W. *J. Chem. Soc., Perkin Trans. 2*, 2000, 1125
[24] Coetzee, J. F., Padmanabhan, G. R. *J. Am. Chem. Soc.*, 1965, *87*, 5005
[25] Kolthoff, I. M., Chantooni, M. K., Bhowmik, S. *J. Am. Chem. Soc.*, 1968, *90*, 23
[26] Augustin-Nowacka, D., Makowski, M., Chmurzynski, L. *Anal. Chim. Acta*, 2000, *418*, 233
[27] Juillard, J., Kolthoff, I. M. *J. Phys. Chem.*, 1971, *75*, 2496
[28] Danil de Namor, A. F., Hill, T. *J. Chem. Soc., Faraday Trans. 1*, 1983, *79*, 2713
[29] Izutsu, K. 'Acid–Base Dissociation Constants in Dipolar Aprotic Solvents' IUPAC Chemical Data Series No. 35, Blackwell, 1990
[30] Coetzee, J. F., Deshmukh, B. K., Liao, C.-C. *Chem. Rev.*, 1990, *90*, 827
[31] Coetzee, J. F., Chang, T. -H., Deshmukh, B. K., Fonong, T. *Electroanalysis*, 1993, *5*, 765
[32] Izutsu, K., Kolthoff, I. M., Fujinaga,T., Hattori, M., Chantooni. M. K. *Anal. Chem.*, 1977, *49*, 503
[33] Zieliska, J., Makowski, M., Maj, K., Liwo, A., Chmurzyski, L. *Anal. Chim. Acta*, 1999, *401*, 317
[34] Pytela, O., Kulhánek, J., Ludwig, M., Řiha, V. *Collect. Czech. Chem. Commum.* 1994, *59*, 627
[35] Barrón, D., Buti, S., Barbosa, J. *Anal. Chim. Acta*, 2000, *403*, 349
[36] Ludwig, M., Baron, V., Kalfus, K., Pytela, O., Veea, M. *Collect. Czech. Chem. Commum.*, 1986, *51*, 2135
[37] Coetzee, J. F., Bertozzi, R. J. *Anal. Chem.*, 1971, *43*, 961
[38] Coetzee, J. F., Bertozzi, R. J. *Anal. Chem.*, 1973, *45*, 1064
[39] Deshmukh, B. K., Siddiqui, S., Coetzee, J. F. *J. Electrochem. Soc.*, 1991, *138*, 124
[40] Barbosa, J., Barrón, D., Bosch, E., Rosés, M. *Anal. Chim. Acta*, 1992, *265*, 157
[41] Barbosa, J., Barrón, D., Bosch, E., Rosés, M. *Anal. Chim. Acta*, 1992, *264*, 229
[42] Streitwieser, A., Ciula, J. C., Krom, J. A., Thiele, G. *J. Org. Chem.*, 1991, *56*, 1074
[43] Streitwieser, A., Wang, D. Z., Stratakis, M., Facchetti, A., Gareyev, R., Abbotto, A., Krom, J. A., Kilway, K. V. *Can. J. Chem.*, 1998, *76*, 765
[44] Streitwieser, A., Facchetti, A., Xie, L., Zhang, X., Wu, E. C. *J. Org. Chem.*, 2012, *77*, 985
[45] Abu-Hasansayn, F., Stratakis, M., Streitwieser, A. *J. Org. Chem.*, 1995, *60*, 4688
[46] Abbotto, A., Streitwieser, A. *J. Am. Chem. Soc.*, 1995, *117*, 6358
[47] Gareyev, R., Ciula, J. C., Streitwieser, A. *J. Org. Chem.*, 1996, *61*, 4589
[48] Streitwieser, A., Kim, Y-J. *J. Am. Chem. Soc.*, 2000, *122*, 11783
[49] Barrón, D., Barbosa, J. *Anal. Chim. Acta*, 2000, *403*, 339
[50] Rodima, T., Kaljurand, I., Pihl, A., Mäemets, V., Leito, I., Koppel, A. *J. Org. Chem.*, 2002, *67*, 1873
[51] Garrido, G., Koort, E., Ràfols, C., Bosch, E., Rodima, T., Leito, I., Rosés, M. *J. Org. Chem.*, 2006, *71*, 9062
[52] Garrido, G., Rosés, M., Ràfols, C., Bosch, E. *J. Soln. Chem.*, 2008, *37*, 689
[53] Abdur-Rashid, K., Fong, T. P, Greaves, B., Gusev, D. G., Hinman, J. G., Landau, S. E., Lough, A. J., Morris, R. H. *J. Amer. Chem. Soc.*, 2000, *122*, 9155

Acid–Base Equilibria and Salt Formation

8

Previous chapters have discussed the influence of solvent on the dissociation of neutral acids, such as carboxylic acids, phenols and carbon acids, and cationic acids, such as protonated amines, anilines, pyridines and phosphazene bases, represented by eqs. (8.1) and (8.2), respectively.

$$HA \rightleftharpoons H^+ + A^- \qquad (8.1)$$

$$BH^+ \rightleftharpoons H^+ + B^- \qquad (8.2)$$

In most instances though, we are primarily interested in *acid–base equilibria* rather than the ionization of individual acids; for example, salt formation between carboxylic acids and nitrogen bases, and the generation of reactive intermediates, such as enolates, in synthetic procedures. We can categorize these equilibria as charge-neutral, eqs. (8.3), (8.4), or charge-forming, eq. (8.5).

$$HA_1 + A_2^- \rightleftharpoons HA_2 + A_1^- \qquad (8.3)$$

$$B_1H^+ + B_2 \rightleftharpoons B_1 + B_2H^+ \qquad (8.4)$$

$$HA + B \rightleftharpoons BH^+ + A^- \qquad (8.5)$$

Importantly, *the solvated proton—a dominant factor in determining the response of individual dissociation constants to solvent—is absent from acid–base equilibria*. One result of this is expected to be a blurring of the distinction between strongly basic and weakly basic aprotic solvents, which we have seen to be very influential in the ionization of individual acids (Chapter 6 and 7).

In principle, the analysis of acid–base equilibria is straightforward, as the equilibrium constants for such reactions follow directly from the ratio of the dissociation constants of the component acids. At the higher concentrations, however, such as typically used in synthetic procedures, association equilibria, including hydrogen-bond association and ion-pair formation between the components, can have a profound effect on the overall equilibria. Solute–solute interactions, reflected in the activity coefficients of the components, also have an increasingly stabilizing effect on ions as the concentrations increase. In this chapter we examine the interacting influence of solvent and reagent concentration on acid–base equilibria.

8.1 Charge-neutral equilibria

For related pairs of acids, such as phenol and 4-nitrophenol, the response to solvent change of charge-neutral equilibria, represented by eqs.(8.3) and (8.4), is essentially a measure of the sensitivity of substituent effects on a given series of acids or bases to solvation.* We have commented upon this separately for different acid–base systems in the various solvents considered during the preceding three chapters, but it is useful briefly to summarize the situation here.

The relationship between the dissociation constant of a substituted acid solvent S, $pK_a(S)$, and that of the (unsubstituted) reference acid, $pK_a^o(S)$, may be represented by eq. (8.6), (eq. (6.22)), in which σ is a constant for a particular substituent, and ρ_S measures the sensitivity of the pK_a to substituent in solvent S.

$$pK_a(S) = pK_a^o(S) - \rho_S(\Sigma\sigma) \qquad (8.6)$$

If we combine this with the analogous equation for dissociations constants in water, $pK_a(H_2O)$, we obtain, after rearrangement,† eq. (8.7), where the constant, $C = pK_a^o(S) - (\rho_S/\rho_w)pK_a^o(H_2O)$, and ρ_w is the observed value of ρ in aqueous solution.

$$pK_a(S) = (\rho_S/\rho_w)pK_a(H_2O) + C \qquad (8.7)$$

Hence, a plot of $pK_a(S)$ against $pK_a(H_2O)$ should have a slope, ρ_S/ρ_w, which reflects the relative sensitivity of the pK_a-values to substituent in the different solvents. Such a plot is shown in Fig. 8.1 for phenols (3,4-substituted) in solvents methanol, dimethylsulphoxide (DMSO) and acetonitrile (MeCN).

The slopes are in all cases greater than one—i.e., the pK_a-values increase more strongly as the acids become weaker—but the effect is noticeably greater for the aprotic solvents (MeCN and DMSO) than for MeOH. The slopes of such plots are very similar across a range of aprotic solvents, including dimethylformamide (DMF), N-methylpyrrolidin-2-one (NMP), dimethylacetamide (DMAC) in addition to DMSO and MeCN. The increases in slope

*Equilibria (8.3) and (8.4) are not normally strongly dependent upon solution concentration: activity coefficients for the charged species largely cancel, and the extent of ion-pair formation and homohydrogen-bond formation for related anions, A_1^- and A_2^-, does not vary strongly with substituent, and therefore does not strongly influence the equilibrium position

†If we write eq. (8.6) in the form $\rho_S(\Sigma\sigma) = \Delta pK_a(S)$, where $\Delta pK_a(S) = pK_a^o(S) - pK_a(S)$, and similarly for aqueous solution, then it follows by division that $\Delta pK_a(S)/\Delta pK_a(H_2O) = \rho_S/\rho_w$

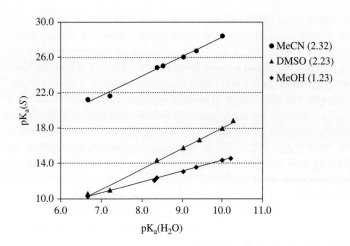

Fig. 8.1. Solvent-dependence of pK_a-values of phenols: the slopes of the lines, corresponding to ρ_S/ρ_w, eq 8.7, are given in parenthesis in the key (pK_a-data from Chapters 5 6,7)

reflect the loss of (H-bond) solvation of the anions on transfer from water and (to a lesser extent) methanol to the aprotic solvents; this is greatest for the most basic anions, which have the highest charge densities.

There is a further increase in slope to 2.95 for phenols in weakly polar tetrahydrofuran (THF) compared with water, consistent with the fact that amongst the aprotic solvents is has the lowest Acceptor Number, and hence by implication the poorest ability to solvate anions.

A similar trend is observed for the carboxylic acids, for which the slopes increase in the order MeOH (1.27) < DMSO (2.36) ~ MeCN (2.32) < THF(4.10).

With respect to equilibria (8.3), it follows that there is an increase in the equilibrium constant for reaction between stronger and weaker acids in all of the solvents compared with water, i.e., an increase the difference between the basicity of the stronger and weaker bases. For example, the equilibrium constant for reaction between phenol and 4-nitrophenoxide in DMSO is more than four orders of magnitude greater than that in water, 1×10^7 compared with 6.3×10^2.

An extreme example of this increase in anion basicity occurs for hydroxide and methoxide anions, with a result that the ionization of weakly acidic substrates, such as anilines, by hydroxide and methoxide, is strongly enhanced in water–DMSO or MeOH–DMSO mixed solvents as the amount of water or methanol decreases (Section 6.1.1).

The influence of solvent on substituent effects is much weaker for cationic acids, such as substituted anilinium and pyridinium ions, eq. (8.4), and is correspondingly less easy to rationalize. Fig. 8.2 shows a plot of dissociation constants for a series of anilines in water against those in methanol and the aprotic solvents DMSO, THF and MeCN. The slopes deviate only modestly from unity and do not follow any obvious trends. Analogous plots for pyridines in MeOH and MeCN have slopes of 0.97 and 1.28, respectively.

We have earlier noted that correlations between pK_a-values among the various solvents, such as those shown in Figs 8.1 and 8.2, which are normally

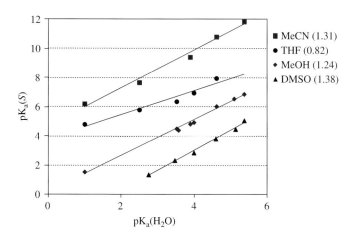

Fig. 8.2.
Solvent-dependence of pK_a-values of anilinium ions: the slopes of the lines, corresponding to $\rho S/\rho_w$, eq 8.7, are given in parenthesis in the key (pK_a- data from Chapters 5–7)

*See Appendix 3.1 for a discussion of composition scales in mixed solvents

8.2 Charge-forming equilibria

8.2.1 Alcohols and mixed-aqueous solvents*

The isolation, and especially the purification, of salts formed between carboxylic acids and biologically active, nitrogen-containing bases is frequently carried out in mixed-aqueous solvents, such as alcohol–water or THF–water mixtures. Such mixtures are also widely used as eluents in chromatography. It is therefore of interest to examine the behaviour of typical carboxylic acids and nitrogen bases in aqueous–solvent mixtures and their resultant equilibria, represented by eq. (8.8), in which $K_e = K_a(RCO_2H)/K_a(R'NH_3^+)$.

$$RCO_2H + R'NH_2 \xrightleftharpoons{K_e} RCO_2^- + R'NH_3^+ \qquad (8.8)$$

In the case of acetic acid and the anilinium ion, the *individual* dissociation constants contributing to eq. (8.8), show quite different responses to solvent composition, as illustrated for EtOH–water mixtures in Fig. 8.3 (Section 5.4) [1, 2].

The behaviour illustrated in Fig. 8.3 is typical of carboxylic acids and protonated amines in a wide variety of aqueous–solvent mixtures. Thus, Figs 8.4 and 8.5 show the change in dissociation constants, ΔpK_a, for acetic acid and aniline with solvent composition in mixtures of water with acetonitrile [3, 4], tetrahydrofuran [5, 6], *iso*-propanol [7], ethanol [1, 2], and methanol [8]; closely similar behaviour is also observed for benzoic acids, phenols and aliphatic amines.

In the case of the carboxylic acids, Fig. 8.4, the monotonic increase in pK_a-values with decreasing water content is due to the decrease in solvation of both the cation and the carboxylate anions, as discussed in earlier chapters, but because of preferential solvation of the ions by water in the mixtures, the most

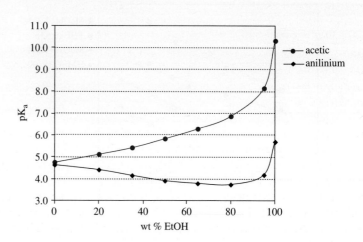

Fig. 8.3.
pK_a-values of acetic acid and anilinium ions in EtOH-water mixtures [1, 2]

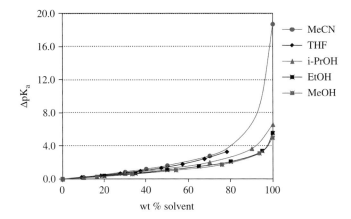

Fig. 8.4.
Change in pK_a-values of acetic acid with composition of aqueous-solvent mixtures [1, 3, 5–8]

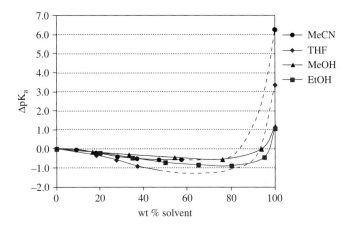

Fig. 8.5.
Change in pK_a-values of the anilinium ion with composition of aqueous-solvent mixtures [2, 4, 5, 8]

rapid increases occur only after the water content of the solvents is severely decreased. In 60 wt% solvent, for example, the increases vary between 1.5 and 2.0 units for all of the mixtures, irrespective of the large differences occurring on transfer to the pure solvents, which range from 5 to 19 pK-units. Very similar trends are observed for acetic acid and benzoic acid in DMF–water and DMSO–water mixtures [9]; for example, the pK_a acetic acid increases by 1.56 units from water to 55 wt% DMF–water compared with a total change of 8.7 units in pure DMF.

Anilinium ions, Fig. 8.5, and ammonium ions in general show somewhat different behaviour. Initially they become more acidic as the organic solvent is added, because (preferential) solvation of the neutral base by the organic component promotes dissociation, and they remain so until the increased energy of the proton dominates at low water content of the solvent.

The changes in the dissociation constants of amines and carboxylic acids with solvent composition (Figs 8.3–8.5), mean that the equilibrium constant for protonation of the amines decreases steadily and significantly with increased organic content of the mixtures, more sharply so as the pure organic solvent is approached. The result is a strongly reduced *intrinsic* tendency towards salt formation between carboxylic acids and amines. In practice, however, the observed behaviour is significantly influenced by the solution concentration, as follows.

In alcohol and alcohol–water mixtures the overall equilibria between carboxylic acids and amines—for example, the protonation of trimethylamine by acetic acid—can be represented by Scheme 8.1. It includes both the proton-transfer equilibrium and ion-pair formation between the resulting ions.

$$HOAc + Me_3N \xrightleftharpoons{K_e} Me_3NH^+ + OAc^- \xrightleftharpoons{K_{IP}} [Me_3NH^+OAc^-]$$

where

$$K_e = K_a(HOAc)/K_a(Me_3NH^+) = \frac{[Me_3NH^+][OAc^-]\gamma_\pm^2}{[HOAc][Me_3N]}$$

$$K_{IP} = \frac{[Me_3NH^+OAc^-]}{[Me_3NH^+][OAc^-]\gamma_\pm^2}$$

Scheme 8.1.
Protonation of trimethylamine by acetic acid in alcohol-water mixtures

In aqueous solution, ion-pair formation is negligible, and the extent of salt formation is largely independent of concentration, apart from a modest increase at higher concentrations due to the decrease in activity coefficient, $\gamma_\pm$. Three things occur as the alcohol content of the solvent increases: the equilibrium constant for protonation decreases; the ion-pair formation constants, K_{IP}, increase;* and the decrease in activity coefficient with concentration becomes much more marked, because of stronger electrostatic interactions in the lower-dielectric media. The former tends to decrease the extent of ionization, whereas the latter two promote increased ionization.

The effect of the combined equilibria in Scheme 8.1 is that for systems in which K_e in water is large, the extent of ionization in pure and mixed alcohol–water remains high *in more concentrated solutions*, despite a significant reduction in K_e. It does, however, fall off precipitately in the non-aqueous media as K_e in water decreases. Thus, although in all cases the equilibrium constant, K_e, reduces strongly and by a similar amount for different acids and amine bases, the influence of solvent on protonation equilibria in concentrated solutions in alcohols and alcohol–water mixtures varies strongly with the value of the equilibrium constant in water.

Table 8.1 presents an analysis of the equilibrium between acetic acid and trimethylamine ($K_e = 1.1 \times 10^5$ in water), in mixtures of ethanol and water at different solution concentrations.† The concentrations chosen include limiting low concentrations (infinite dilution), 0.01M and 0.25M; the latter value is representative of concentrations used for reaction and salt isolation, whereas 0.01M is more typical of concentrations used in the determination of dissociation constants.

The most striking feature of the results reported in Table 8.1 is the increase in the extent of protonation of the free base with concentration in pure ethanol, from a limiting low-concentration value of 37% to around 85% at

*The ion-pair formation constants of, for example, $[Bu_4N^+Br^-]$ in MeOH, EtOH, i-PrOH, and t-BuOH are 59, 224, 1.29×10^3 and 1.17×10^5 M^{-1}, respectively (Rosés, M., Rived, F., Bosch. E. *J. Chem. Soc., Faraday Trans. 1,* 1993, *89*, 1723).

†Equilibria in Scheme 8.1 were solved numerically using Micromath Scientist™ and data from Chapter 5, refs. [1, 2] and Rosés, M., Rived, F., Bosch. E. *J. Chem. Soc., Faraday Trans. 1,* 1993, *89*, 1723); K_{IP}-values are representative of typical electrolytes in the solvents, derived from conductivity studies

Table 8.1 Equilibria between acetic acid and trimethylamine (Scheme 8.1) in water-ethanol mixtures at 25°C

Wt% EtOH[a]	[HOAc] = [Me$_3$N/M	$\gamma_\pm^b$	%Me$_3$N	%Me$_3$NH$^+$	% Ion-pair	% Ionization[c]
0	0[d]	1	0.31	99.7	0	99.7
0	0.01	0.90	0.28	99.7	0	99.7
0	0.25	0.75	0.24	99.8	0	99.8
65	0[d]	1	8.5	91.5	0	91.5
65	0.01	0.79	6.2	84.7	9.0	93.7
65	0.25	0.51	2.8	55.6	41.6	97.2
100	0[d]	1	63.2	36.8	0	36.8
100	0.01	0.69	46.9	39.0	14.0	53.0
100	0.25	0.25	15.2	38.6	46.2	84.8

[a] Values for K$_e$, K$_{IP}$/M^{-1} and the Debye-Hückel A-factor in the mixtures are, respectively: water, 1.1 × 10^5, 0, 0.51; 65 wt% EtOH, 1.2 × 10^2, 20, 1.24; 100% EtOH, 0.4, 200, 2.90; [c] Mean ionic activity coefficient, Chapter 4, Section 4.2; [c] % Me$_3$NH$^+$ + %[Me$_3$NH$^+$OAc$^-$]; [d] Limiting values at low solution concentration

a concentration of 0.25 M, despite the small equilibrium constant for the protonation reaction. It is noticeable that the major contributor to this increase is the extensive ion-pair formation, representing 46% of the total amount of acetic acid and trimethylamine initially present. In water, by contrast the extent of ion-pair formation is negligible, but salt formation in solution is almost quantitative at all concentrations, because of the large difference in pK$_a$-values for acetic acid and the trimethylamonium ion, which corresponds to an equilibrium constant for protonation of trimethylamine by acetic acid of K$_e$ = 1.1 × 10^5. The reduced activity coefficient, $\gamma_\pm$, also contributes to a stabilization of the ions in ethanol and 65% ethanol–water. The net result is that the extent of protonation of trimethylamine by acetic acid remains at 85% or greater across the whole range of solvent compositions, despite a reduction in the equilibrium constant for protonation of more than five orders of magnitude.

The behaviour exhibited in Table 8.1 is typical of the protonation of *strongly basic amines* by aliphatic and aromatic carboxylic acids in ethanol and similarly in methanol and *iso*-propanol, as shown in Table 8.2. At the higher concentration of 0.25 M, more than 80% overall protonation is achieved in each of the pure solvents and in their mixtures with water.

In each case, stabilization of the ions at higher concentrations by electrostatic attractions, as measured by the activity coefficient, $\gamma_\pm$, and especially by ion-pair formation, leads to high overall levels of ionization, despite a low intrinsic equilibrium constant for the primary ionization reaction.

The outcome for equilibria of the type described in Scheme 8.1 on transfer from water to alcohols and aqueous–alcohol mixtures, however, changes strongly as ΔpK$_a$ (and hence log K$_e$) in water decreases. This is because, as the basicity of the amine decreases and hence the equilibrium constant for protonation becomes smaller, the overall extent of ionization is decreased by a combination of two reinforcing factors: a strongly reduced tendency to form the free ions, and a resultant diminished extent of ion-pair formation, even at high overall solution concentrations.

Table 8.2 Equilibria between acetic acid and trimethylamine (Scheme 8.1) in water and alcohols at 25°C

Solvent[a]	[HOAc] = [Me$_3$N/M	$\gamma_\pm^b$	% Me$_3$N	% Me$_3$NH$^+$	% Ion-pair	% Ionization[c]
Water	0[d]	1	0.31	99.7	0	99.7
	0.25	0.75	0.24	99.8	0	99.8
MeOH	0[d]	1	47.7	52.3	0	52.3
	0.25	0.39	15.2	42.7	42.0	84.7
EtOH[e]	0[d]	1	63.2	36.8	0	36.8
	0.25	0.25	15.2	38.6	46.2	84.8
i-PrOH	0[d]	1	65.5	34.5	0	34.5
	0.25	0.10	7.4	38.8	53.8	92.6

[a] Values for K$_e$, K$_{IP}$/M^{-1} and the Debye-Hückel A-factor in MeOH and i-PrOH are, respectively: MeOH, 1.2, 60, 1.9; i-PrOH, 0.3, 1300, 4.7; [b] Mean ionic activity coefficient, Chapter 04, Section 4.2; [c] % Me$_3$NH$^+$ + %[Me$_3$NH$^+$OAc$^-$]; [d] Limiting values at low solution concentration; [e] Table 8.1

The effect can be quite dramatic, as shown by the results in Table 8.3 for a comparison of reactions of acetic acid (aqueous pK$_a$ = 4.76) with trimethylamine (aqueous pK$_a$ = 9.80, ΔpK$_a$ = 5.04), N-methyl morpholine (aqueous pK$_a$ = 7.60, ΔpK$_a$ = 2.84), and N, N-dimethylaniline (aqueous pK$_a$ = 5.07, ΔpK$_a$ = 0.31) at a solution concentration of 0.25M in water and ethanol.

The exact details will depend upon the magnitude of the ion-pair formation constants for a given system, but the general trend is clear. As the equilibrium constant in water becomes smaller, the equilibrium position in ethanol (and other alcohols) becomes increasingly unfavourable, and very low levels of ionization occur, despite a potentially strong tendency towards ion-pair formation.

8.2.2 Polar aprotic solvents

The equilibrium between carboxylic acids and amines in polar aprotic solvents is not strongly sensitive to the solvent basicity, because the solvated proton is

Table 8.3 Equilibria between acetic acid and trimethylamine, N-methylmorpholine, and N, N-dimethylaniline (Scheme 8.1) in water and ethanol at 25°C[a]

Base	Solvent	$\gamma_\pm^b$	%BH$^+$	% Ion-pair	% Ionization[c]
trimethylamine	H$_2$O	0.75	99.8	0	99.8
	EtOH[d]	0.25	38.6	46.2	84.8
N-Me-morpholine	H$_2$O	0.75	97.4	0	97.4
	EtOH[d]	0.70	9.3	21.1	30.4
N, N-Me$_2$-aniline	H$_2$O	0.76	64.8	0	64.8
	EtOH[d]	0.87	1.1	0.4	1.6

[a] Solution concentration = 0.25M; [b] Mean ionic activity coefficient, Section 4.2, A = 0.509 (H$_2$O). 2.90 (EtOH); [c] % BH$^+$ + % [BH$^+$OAc$^-$]; [d] Values used for K$_e$ and K$_{IP}$/M^{-1} for Me$_3$N, N-methylmorpholine and N,N-dimethylaniline in ethanol are, respectively: 0.4 and 200; 8.7 × 10^{-3} and 200; 1 × 10^{-4} and 200

Table 8.4 Solvent dependence of the equilibrium constants for reaction between acetic acid and triethylamine at 25°C[a]

Solvent	H$_2$O	MeOH	DMSO	NMP	DMF	DMAC	MeCN
LogK$_e$[b]	5.92	1.06	−3.60	−4.80	−4.25	−3.50	−5.00

[a] Solvent abbreviations as in Table 1.1; [b] Eq 8.9, $K_e = K_a(CH_3CO_2H)/K_a(Et_3NH^+)$, data from Chapter 6 and 7

no longer involved. The magnitude of the equilibrium constant for the reaction between acetic acid and triethylamine, eq. (8.9) is shown for several aprotic solvents in Table 8.4, which also includes results for the protic solvents water and methanol for comparison.

$$CH_3CO_2H + Et_3N \xrightleftharpoons{K_e} Et_3NH^+ + CH_3CO_2^- \qquad (8.9)$$

Neglecting the relatively small differences between the aprotic solvents, the important feature of the data is that the equilibrium in aprotic solvents constant is universally low, $K_e \sim 10^{-4}$. The consequence is that the concentration of free ions will *at most* be around 1% of the total concentration of acetic acid and triethylamine, and of course will be even lower for less basic amines.

The low concentration of free ions, combined with the modest ion-pair constants in these polar solvents ($K_{IP} \sim 50 - 200 M^{-1}$) [10–12], also means that ion-pair formation between Et$_3$NH$^+$ and CH$_3$CO$_2^-$ will make little contribution to the overall equilibrium. We do, however, need to take into consideration homohydrogen-bond formation between acetic acid and the acetate ion in the overall equilibration process, which can be represented by Scheme 8.2.

$$HOAc + Et_3N \xrightleftharpoons{K_e} Et_3NH^+ + OAc^-$$
$$HOAc + OAc^- \xrightleftharpoons{K_{AHA}} [AcO\cdots HOAc]^-$$
$$Et_3NH^+ + [AcO\cdots HOAc]^- \xrightleftharpoons{K_{IP}} \{Et_3NH^+[AcO\cdots HOAc]^-\}$$

Scheme 8.2.
Protonation of triethylamine by acetic acid in aprotic solvents

In principle, homohydrogen-bond association involving Et$_3$NH$^+$ and Et$_3$N can occur, but the equilibrium constants are very small, even in acetonitrile [13].

Table 8.5 shows an analysis of the composition of an equilibrium mixture of acetic acid and triethylamine in dimethylsulphoxide and acetonitrile at a solution concentration of 0.25M.*

The results for the two solvents are similar, and would reflect those for other polar aprotic solvents. In each cases the level of free acetate is very low, but in acetonitrile the overall equilibrium is shifted modestly towards a greater degree of ionization because of the higher homohydrogen-bond formation constant: $K_{AHA} = 4.7 \times 10^3 M^{-1}$ in acetonitrile, compared with $K_{AHA} = 30 M^{-1}$ in dimethylsulphoxide [14]. Protonation of more weakly basic amines, and anilines and pyridines, will be negligible in these solvents.

*Equilibria in Scheme 8.2 were solved numerically using Micromath Scientist^{TM} using the data from the footnotes in Table 8.5

Table 8.5 Equilibria between acetic acid and triethylamine in dimethylsulfoxide and acetonitrile (Scheme 8.2) at 25°C[a]

Solvent	$\gamma_{\pm}^{b}$	%$CH_3CO_2^-$	%Et_3N	%Et_3NH^+	%$[AHA]^-$[c]	% Ion-pair	% Ionization[d]
DMSO[e]	0.80	0.6	90.0	4.5	3.8	5.5	9.3
MeCN[f]	0.65	0.01	77.8	8.1	8.1	14.1	22.2

[a] Solution concentration = 0.25M; [b] Mean ionic activity coefficient, Chapter 4, Section 4.2, A = 1.04 (DMSO). 1.55 (MeCN); [c] $[AHA]^- = [AcO\cdots\cdots HOAc]$; [d] % ionization = % Et_3NH^+ + %ion − pair; [e] $K_e = 2.5 \times 10^{-4}$, $K_{AHA} = 30M^{-1}$, $K_{IP} = 200M^{-1}$; [f] $K_e = 1.0 \times ^{-5}$, $K_{AHA} = 4700M^{-1}$, $K_{IP} = 200M^{-1}$

It is of interest also to look at the equilibria between the strongest of the neutral bases, the phosphazene bases [15–17], and simple carbon acids, such as acetone and acetophenone, in view of their increasing use in synthetic procedures [18]. Amongst the strongest of these is the triphosphazene, $PhP_3(dma)$, the conjugate acid of which has a $pK_a = 31.5$ in acetonitrile [17], which would correspond to $pK_a = 22.0$ in dimethylsulphoxide (Chapters 6 and 7).

$PhP_3(dma)$

Combining the above $pK_a(PhP_3(dma)H^+)$ with those of acetone ($pK_a = 26.5$) and acetophenone ($pK_a = 24.7$) in dimethylsulphoxide [19], leads to the equilibrium distribution shown in Table 8.6, for a solution concentration of 0.25M. The systems may be represented by Scheme 8.1, as no significant homohydrogen-bond formation accompanies the ionization of carbon acids.

It is clear from the data in Table 8.6 that, even for this very strong phosphazene base, small amounts only of the enolate ions occur at equilibrium, and the same would be true in other polar aprotic solvents; in practice, in reported synthetic procedures the enolate is trapped *in situ* as a reactive intermediate, using the highly active sulphonyl fluoride, $CF_3(CF_2)_2SO_2F$, for example, and the results in Table 8.6 explain the need for this. Quantitative generation of the enolate ions for these unactivated ketones requires use of a much stronger base, such as lithium diisopropylamide [20].

Table 8.6 Equilibria between acetone, acetophenone, and phosphazene $PhP_3(dma)$ in dimethylsulfoxide (Scheme 8.1) at 25°C[a]

Ketone	$\gamma_{\pm}^{b}$	% Enolate	%$PhP_3(dma)$	% Ion-pair	% Ionization[c]
Acetone[d]	0.91	0.6	99.2	0.2	0.8
Acetophenone[e]	0.79	4.9	87.4	7.6	12.5

[a] Solution concentration = 0.25M; [b] Mean ionic activity coefficient, Ch 4, Section 4.2, A = 1.04; [c] % ionization = % enolate + % ion-pair; [d] $K_e = 3.2 \times 10^{-5}$, $K_{IP} = 100M^{-2}$; [e] $K_e = 2.0 \times 10^{-3}$, $K_{IP} = 100M^{-2}$

More highly activated carbon acids, such as nitroalkanes, ketoesters and diketones (Table 6.6), would, however, be readily ionized by the strong phosphazene bases in aprotic solvents.

8.2.3 Non-polar aprotic solvents

We have seen that equilibria between simple amines and carboxylic acids (and phenols) in *polar* aprotic solvents lie predominantly in favour of the unionized species for a combination of two reasons: intrinsic equilibrium constants are low, primarily because of the very poor solvation of the carboxylate anions; and the relatively high polarity of the solvents means that association equilibria, such as ion-pair and homohydrogen-bond formation, are not sufficiently high to compensate for this.

In solvents of very low polarity, however, whilst there is no observable formation of free ions, a strong degree of ionization to form ion-pairs is frequently observed. In chlorobenzene, for example, the increased absorbance of the anion of acid–base indicators, such as 2, 6-dinitrophenol, in the presence of amines or anilines suggests high levels of ion-pair formation. Thus, the equilibrium constant for ion-pair formation from reaction between tributylamine (B) and 2, 6-dinitrophenol (HA), eq. (8.10), is very large [21]: $K_I = 6.9 \times 10^4 M^{-1}$.

$$\text{ArOH} + \text{Bu}_3\text{N} \xrightleftharpoons{K_I} [\text{Bu}_3\text{N}^+\text{ArO}^-] \quad (8.10)$$

Absorption studies in the visible or UV region, such as that used to quantify eq. (8.10), cannot give structural details of the ion-pairs, but IR spectroscopy has been used to good effect in this respect. Barrow and Yeager have used detailed IR studies to elucidate the nature of the ion-pair species formed in the reaction of acetic acid with tertiary, secondary and primary ethylamine in carbon tetrachloride and chloroform, and to determine the corresponding equilibrium constants [22–24].

In titrations of acetic with Et_3N, performed at a range of concentrations up to 0.3M, it was found that at half neutralization in carbon tetrachloride there was a complete loss of IR stretches characteristic of acetic acid monomers and dimers with formation of a species of stoichiometry $Et_3N(HOAc)_2$ [22]; the resulting species was identified as I. Addition of further quantities of Et_3N led to the disappearance of the intermediate I and formation of ion-pair II. The equilibrium constant for formation of II has a value $K_e = [II]/[HOAc][Et_3N] = 800 M^{-1}$.

A similar result was observed in the hydrogen-bonding solvent, chloroform, but in this case the stoichiometric ion-pair involved additional hydrogen-bonding between the ion-pair and a molecule of chloroform, as in III: $K_e = [III]/[HOAc][Et_3N] = 3000 M^{-1}$.

$$\text{H}_3\text{C}-\underset{\underset{\text{O}\cdots\text{HNEt}_3}{+}}{\overset{\text{O}\cdots\text{HCCl}_3}{\diagdown}}-$$

III

The same general behaviour was found for equilibria involving the secondary and primary amines, Et$_2$NH and EtNH$_2$, respectively [23, 24]. There was also in each case a complete loss of free NH-stretches in the ion-pairs, so that both (or all three) N–H hydrogens were involved in H-bond formation with the carboxylate anion. Very high ion-pair formation constants, K_I (eq. (8.10)) have also been reported for reaction between butylamine and a wide range of phenols and carboxylic acids [25]: log K_I values ranged from 1.9 for 2,5-dinitrophenol to 7.05 for trichloroacetic acid.

It is likely that the ionization to form ion-pairs between amines and carboxylic acids or phenols will be quite general in aprotic solvents of *very low* polarity. Ion-pair equilibria between phosphazene bases and C–H and N–H acids in heptane ($\varepsilon_r = 1.92$), for example, have also been reported [26]. In such solvents, the low dielectric constants mean that energy derived from electrostatic attraction between the ions, combined with additional hydrogen-bond stabilization (e.g., II above), is very high. At low concentrations (below K_I^{-1}M), of course, entropy predominates and the systems revert increasingly to the free acids and bases.

Finally, we mention the important solvent tetrahydrofuran, which is of low but intermediate polarity ($\varepsilon_r = 7.6$) with respect to the extremes of high- and low-polarity solvents discussed above. This solvent has been fully discussed in Chapter 7, but we note here for completeness that by working at very low solution concentrations (typically $< 10^{-5}$M), and making allowance for ion-pair and ion-triplet formation [27–31], it has been possible to derive conventional dissociation constants. At any useful practical concentration, however, the fraction of free ions is negligible. This has led to the derivation of a set of ion-pair acidities [32, 33], based on either the the caesium cation, which forms contact ion-pairs, as in eq. (8.11), or the lithium cation, which tends to form solvent-separated ion-pairs in which the lithium ion retains a full coordination sphere of THF molecules.

$$\text{RH} + \text{Cs}^+\text{R}'^- \xrightleftharpoons{K} \text{Cs}^+\text{R}^- + \text{R}'\text{H} \qquad (8.11)$$

References

[1] Gutbezahl, B., Grunwald, E. *J. Am. Chem. Soc.,* 1953, *75,* 559
[2] Grunwald, E., Berkowitz, B. J. *J. Am. Chem. Soc.,* 1951, *73,* 4939
[3] Barbosa, J., Beltrán, J. L., Sanz-Nebot, V. *Anal. Chim. Acta,* 1994, *288,* 271
[4] Espinosa, S., Bosch, E., Rosés, M. *Anal. Chim. Acta,* 2002, *454,* 157
[5] Reynard, R. *Bull. Soc. Chim. Fr.,* 1967, 4597
[6] Muinasamaa, U., Ràfols, C., Bosch, E., Rosés, M. *Anal. Chim. Acta,* 1997, *340,* 133
[7] Bosch, E., Ràfols, C., Rosés, M. *Anal. Chim. Acta,* 1995, *302,* 109
[8] Barcarella, A. J., Grunwald, E., Marshall, H.P., Purlee, E. L. *J. Org. Chem.,* 1955, *20,* 747
[9] Azab, H. A., Ahmed, L. T., Mahmoud, M. R. *J. Chem. Eng. Data,* 1995, *40,* 523

[10] Olmstead, W. N., Bordwell, F. G. *J. Org. Chem.,* 1980, *45*, 3299
[11] Chantooni, M. K., Kolthoff, I. M. *J. Am. Chem. Soc.,* 1968, *90*, 3005
[12] Roland, G., Chantooni, M. K., Kolthoff, I. M. *J. Chem. Eng. Data,* 1983, *28*, 162
[13] Coetzee, J. F., Padmanabhan, G. R. *J. Am. Chem. Soc.,* 1965, *87*, 5005
[14] Chantooni, M. K., Kolthoff, I. M. *J. Phys. Chem.,* 1973, *77*, 527
[15] Kaljurand, I., Kütt, A., Sooväli, L., Rodima, T., Mäemets, V., Leito, I., Koppel, I. A. *J. Org. Chem.,* 2005, *70*, 1019
[16] Schwesinger, R., Willaredt, J., Schlemper, H., Keller, M., Schmitt, D., Fritz, H. *Chem. Ber.* 1994, *127*, 2435 and references therin
[17] Kaljurand, I., Rodima, T., Leito, I., Koppel, I. A., Schwesinger, R. *J. Org. Chem.,* 2000, *65*, 6202
[18] Lyapkalo, I. M., Vogel, M. A. K. *Angew. Chem. Int. Ed.,* 2006, *45*, 4019
[19] Bordwell, F. G. *Acc. Chem. Res.,* 1988, *21*, 456, and references therein
[20] Larock, R. C. 'Comprehensive Organic Transformations', VCH, NY, 1989, p738
[21] Bell, R. P., Bayles, J. W. *J. Chem. Soc.,* 1952, 1518
[22] Barrow, G. M., Yerger, E. A. *J. Am. Chem. Soc.,* 1954, *76*, 5211
[23] Yerger, E. A., Barrow, G. M. *J. Am. Chem. Soc.,* 1955, *77*, 4474
[24] Yerger, E. A., Barrow, G. M. *J. Am. Chem. Soc.,* 1955, *77*, 6206
[25] Rumeau, M., Tremillon, B. *Bull. Chim. Soc. Fr.,* 1964, 1049
[26] Rõõm, E. I., Kaljurand, I., Leito, I., Rodima, T., Koppel, I. A., Vlaslov, V. M. *J. Org. Chem.,* 2003, *68*, 7795
[27] Barrón, D., Barosa, J. *Anal. Chim. Acta*, 2000, *403*, 339
[28] Rodima, T., Kaljurand, I., Pihl, A., Mäemets, V., Leito, I., Koppel, A. *J. Org. Chem.,* 2002, *67*, 1873
[29] Garrido, G., Koort, E., Ràfols, C., Bosch, E., Rodima, T., Leito, I., Rosés, M. *J. Org. Chem.,* 2006, *71*, 9062
[30] Garrido, G., Rosés, M., Ràfols, C., Bosch, E. *J. Soln. Chem.,* 2008, *37*, 689
[31] Barbosa, J., Barrón, D., Bosch, E., Rosés, M. *Anal. Chim. Acta,* 1992, *264*, 229
[32] Streitwieser, A., Ciula, J. C., Krom, J. A., Thiele, G. *J. Org. Chem.,* 1991, *56*, 1074
[33] Streitwieser, A., Wang, D. Z., Stratakis, M., Facchetti, A., Gareyev, R., Abbotto, A., Krom, J. A., Kilway, K. V. *Can. J. Chem.,* 1998, *76*, 765

9 Appendices: Dissociation Constants in Methanol and Aprotic Solvents

9.1 Methanol

The most comprehensive review of data is: Rived, F., Rosés, M., Bosch, E. *Anal. Chim. Acta*, 1998, *374*, 309. Dissociation constants reported below are from this reference unless otherwise stated.

9.1.1 Carboxylic acids and phenols

Acid		pK_a	Acid	pK_a	Acid	pK_a
Carboxylic acid:			**Benzoic**		2-methoxy	9.26
Aliphatic			2,6-dinitro	6.30	2-methoxy	9.30
$CHCl_2CO_2H$		6.38	2,4-dinitro	6.45	4-methoxy	9.79
$CH(CN)CO_2H$		7.50	3,5-dinitro	7.38	2-methyl	9.24
CH_2ClCO_2H		7.88	3,4-dinitro	7.44	3-methyl	9.39
$CH_3CHClCO_2H$		8.06	4-chloro,3-nitro	8.10	4-methyl	9.51
CH_2FCO_2H		7.99	2,6-dichloro	7.05	3-hydroxy	9.58
CH_2BrCO_2H		8.06	3,5-dichloro	8.26	4-hydroxy	9.99
CH_2ICO_2H		8.38	3,4-dichloro	8.53	4-amino	10.25
$CH_2(OH)CO_2H$		8.68	2-nitro	7.64		
$PhCH_2CO_2H$		9.34	3-nitro	8.32	**Phenol:**	
CH_3CO_2H		9.63	4-nitro	8.34	2,4,6-trinitro	3.55
$CH_3CH_2CO_2H$		9.71	2-fluoro	8.41	2,6-dinitro	7.64
$CH_3(CH_2)_2CO_2H$		9.69	3-fluoro	8.87	2,4-dinitro	7.83
oxalic:[a]	pK_{a1}	6.10	4-fluoro	9.23	2,5-dinitro	8.94
	pK_{a2}	10.7	2-chloro	8.31	3,5-dinitro	10.29
maleic:[b]	pK_{a1}	5.7	3-chloro	8.83	3,5-dichloro	12.11
	pK_{a2}	12.8	4-chloro	9.09	2-nitro	11.53
malonic:[a]	pK_{a1}	7.50	2-bromo	8.19	3-nitro	12.41
	pK_{a2}	12.4	3-bromo	8.80	2-chloro	12.97
fumaric:[b]	pK_{a1}	7.9	4-bromo	8.93	3-chloro	13.10
	pK_{a2}	10.3	2-iodo	8.24	4-chloro	13.59
succinic:[a]	pK_{a1}	9.10	3-iodo	8.89	3-bromo	13.30
	pK_{a2}	11.5	4-iodo	9.04	4-bromo	13.63
glutamic:[a]	pK_{a1}	9.40	3-methylsulphonyl	8.43	H	14.33
	pK_{a2}	11.5	3-cyano	8.53	2-methoxy	14.48

Adipic:[a]	pK$_{a1}$	9.45	4-cyano	8.42	2-methyl	14.86
	pK$_{a2}$	11.1	3-acetyl	8.87	3-methyl	14.43
			4-acetyl	8.72	4-methyl	14.54
			1-H	9.30	4-t-butyl	14.52

[a] Chantooni, M. K.; Kolthoff, I. M. *J. Phys. Chem.*, 1975, *79*, 1176; [b] Kolthoff, I. M., Chantooni, M. K. *Anal. Chem.*, 1978, 50, 1440; Garrido, G.; de Nogales, V.; Ràfols, C, Bosch, E. *Talanta*, 2007, *73*, 115

9.1.2 Protonated nitrogen bases

Base	pK$_a$	Base	pK$_a$	Base	pK$_a$
Amine:		3-fluoro	4.60	2-bromo	1.0
hydroxylamine	6.29	2-chloro	3.71	3-bromo	2.90
ammonia	10.78	3-chloro	4.52	3-cyano	1.4
ethanolamine	10.88	4-chloro	4.95	4-cyano	1.9
methylamine	11.00	2-bromo	3.46	2-acetyl	3.54
dimethylamine	11.20	3-bromo	4.42	3-acetyl	3.73
trimethylamine	9.80	4-bromo	4.84	2-hydroxy	2.79
ethylamine	11.00	H	6.05	3-hydroxy	5.74
triethylamine	10.78	2-methyl	5.95	H	5.44
butylamine	11.48	3-methyl	6.09	2-methyl	6.18
cyclohexylamine	11.68	4-methyl	6.57	3-methyl	6.04
piperidine	11.07	3-methoxy	6.04	4-methyl	6.40
NMe$_4$-guanidine	13.20	4-methoxy	6.89	3,5-dimethyl	6.43
				3,4-dimethyl	6.83
Aniline:		**Pyridine:**[a]		2,6-dimethyl	6.86
2-nitro	0.20	2-chloro	1.0	4-dimethylamino	10.10
4-nitro	1.55	3-chloro	2.83	4-amino	10.37

[a] Augustin-Nowacka, D., Malowski, M.; Chmurzynski, L. *Anal. Chim. Acta*, 2000, *418*, 233

Additional dissociation constants for neutral acids and cationic acids in methanol may be estimated from correlations with those in water, as follows:

$$pK_a(MeOH) = m \, pK_a(H_2O) + c$$

The best-fit values of m and c for the different acid types are:

Acid type	m	c
Carboxylic acids	1.02	4.98[a]
Phenols	1.08	3.66
Protonated nitrogen bases[b]	1.02	0.72

[a] For the second pK$_a$ of dicarboxylic acids, $c = 6.2$; [b] a correlation based on anilines alone gives $m = 1.21$ and $c = 0.38$

9.2 Dimethylsulphoxide

Data comes from Bordwell, F.G. *Acc. Chem. Res.*, 1988, 21, 456, and references therein, unless otherwise indicated. A comprehensive listing is given on the Web, Bordwell pK_a Table (Acidity in DMSO), © 2001-20011 Hans J. Reich.

9.2.1 Carboxylic acids, alcohols, phenols

Acid		pK_a	Acid	pK_a	Acid	pK_a
Carboxylic acid:			$3,4-Cl_2$	11.0	$3-NO_2$	14.4
Aliphatic[a]			2-Cl	11.2	$4-NMe_3^+$	14.7
$CHCl_2CO_2H$		6.4	3-Br	11.3	2-CN	12.1
CH_2ClCO_2H		8.9	4-Cl	11.5	4-CN	13.2
CH_3CO_2H		12.6	4-Br	11.6	3-CN	14.8
$CH_3(CH_2)_2CO_2H$		12.9	H	12.4	$4-SO_2Me$	13.6
oxalic:[b]	pK_{a1}	6.2	3-Me	12.4	$4-CF_3$	15.3
	pK_{a2}	14.9	3-OH	12.5	$3-CF_3$	15.6
malonic:[b]	pK_{a1}	6.9	4-Me	12.6	3-Cl	15.8
	pK_{a2}	18.5	$3,4-Me_2$	13.0	4-Cl	16.7
succinic:[b]	pK_{a1}	9.5	$4-NH_2$	14.0	2-F	15.6
	pK_{a2}	16.5			4-F	18.0
glutaric:[b]	pK_{a1}	10.9	**Alcohol:**		H	18.0
	pK_{a2}	15.3	HOH	31.4	$2-NH_2$	18.2
adipic:[b]	pK_{a1}	11.9	MeOH	29.0	2-OMe	17.8
	pK_{a2}	14.1	EtOH	29.8	4-OMe	19.1
o-phthalic[b]:	pK_{a1}	5.9	i-PrOH	30.3	4-Me	18.9
	pK_{a2}	16.0	t-BuOH	32.2	$4-NMe_2$	19.8
			CF_3CH_2OH	23.5		
Benzoic[a]			$(CF_3)_2CHOH$	17.9	**Thiophenol:**	
$2,6-(OH)_2$		3.1	$(CF_3)_3COH$	10.7	$4-NO_2$	5.5
$2,4-(NO_2)_2$		5.2			2-chloro	8.55
2-OH		8.2	**Phenol:**		3-chloro	8.57
$3,5-(NO_2)_2$		8.8	$2,6-(NO_2)_2$	6.2	4-bromo	8.98
$2-NO_2$		9.9	$2,4-(NO_2)_2$	6.3	H	10.3
$4-Cl, 3-NO_2$		10.0	$3-CF_3, 4-NO_2$	10.4	3-Me	10.55
$3,5-Cl_2$		10.4	$3,5-(NO_2)_2$	11.3	4-OMe	11.19
$4-NO_2$		10.6	$4-NO_2$	12.2		
$3-NO_2$		10.8	$2-NO_2$	12.2		

[a] Ref: Ritchie, C. D., Uschold, R. E. *J. Am. Chem. Soc.*,1967, 89, 1721; 1968, 90, 2821: Kolthoff, I. M.; Chantooni, M. K. *J. Phys. Chem.*, 1971, 93, 3843; Pytela, O., Kuhlánek, J., Ludwig, M., Řiha, V. *Collect. Czech. Chem., Commun.*, 1994, 59, 627; Pytela, O., Kulhánek, J. *Collect. Czech. Chem. Commun.* 1997, 62, 913; [b] Chantooni, M. K., Kolthoff, I. M. *J. Phys. Chem.*, 1975, 79, 1176

Further values may be estimated from aqueous data through correlations between dissociation constants in water and DMSO (Chapter 6) as follows.

Carboxylic acids $pK_a(DMSO) = 1.57 pK_a(H_2O) + 4.21$
Phenols $pK_a(DMSO) = 1.98 pK_a(H_2O) - 2.40$
Thiophenols $pK_a(DMSO) = 2.55 pK_a(H_2O) - 6.26$

9.2.2 Inorganic acids and miscellaneous

Acid	pK_a	Acid	pK_a	Acid	pK_a
HBr	0.9	HN_3	7.9	HF	15
CH_3SO_3H	1.6	NH_4^+	10.5	NH_2CN	16.9
HCl	1.8	HSO_4^-	12.6[a]	H_2O	32
CF_3CO_2H	3.4	HCN	12.9		
HNO_2	7.5				

[a] Kolthoff, I. M., Chantooni, M. K. *J. Am. Chem. Soc.*, 1968, 90, 5961

9.2.3 Anilines, anilides, amides (N–H-ionization)

Acid	pK_a	Acid	pK_a	Acid	pK_a
Anilines[a]		4-Br	29.1	R = Me; R′ = Me	25.9
2, 4–$(NO_2)_2$	15.9	H	30.7		
4–NO_2	20.9	3-Me	31.0		
2, 6–Cl_2	24.8	**Amides**[b]		**Anilides**[c]	
4–$COCH_3$	25.3	R = CF_3; R′ = H	17.2	X = Br; Y = CN	15.4
4-CN	25.3	R = 4–Py; R′ = H	21.6	X = Br; Y = Cl	17.0
2, 4–Cl_2	26.3	R = 3–Py; R′ = H	22.0	X = Br; Y = H	18.0
3–CF_3	28.5	R = Ph; R′ = H	23.3	X = H; Y = CN	19.0
3-Cl	28.5	R = H; R′ = H	23.5	X = H; Y = Cl	20.7
4-Cl	29.4	R = Me; R′ = H	25.5	X = H; Y = H	21.5

[a] Bordwell, F. G., Algrim, D. J. *J. Am. Chem. Soc.*, 1988, 110, 2964; [b] Hansen, M. M., Harkness, A. R., Coffey, D. S., Bordwell, F. G., Zhao, Y. *Tett. Letters*, 1995, 36, 8949; [c] Bordwell, F. G., Gou-Zhen, J. *J. Am. Chem. Soc.*, 1991, 113, 8398

9.2.4 Carbon acids: ketones, esters, nitroalkanes

Acid[a]	pK_a	Acid[a]	pK_a	Acid[a]	pK_a
Ketones: $H_3C-CO-CH_2X$		$X-C_6H_4-CO-CH_3$		**Esters:** $EtO-CO-CH_2X$	
X = SO_2Ph	12.5	X = 4-CN	22.0	X = NO_2	9.1
X = $COCH_3$	13.3	X = 4–CF_3	22.7	X = CN	12.5
X = COPh	14.2	X = 3–CF_3	22.8	X = $COCH_3$	14.2
X = SOPh	15.1	X = 3-Cl	23.2	X = CO_2Et	16.4
X = Ph	19.8	X = 4-Cl	23.8	X = Ph	22.7
X = H[a]	26.5[b]	X = 4-Br	23.8	X = H	29.5
Ph–CO–CH_2X		X = 3-F	23.5	**Nitroalkanes:** NO_2CH_2X	
X = NO_2	7.7	X = 4-F	24.5	X = NO_2	6.6
X = CN	10.2	X = H	24.7	X = SO_2Ph	7.1

Acid	pKa	Acid	pKa	Acid	pKa
X = SO$_2$Ph	11.4	X = 3-OMe	24.5	X = COPh	7.7
X = COPh	13.4	X = 4-OMe	25.7	X = CO$_2$Et	9.1
X = Ph	17.7	X = 3-NMe$_2$	25.3	X = Ph	12.1
X = F	21.7	X = 4-NMe$_2$	27.5	X = Et	17.0
X = OMe	22.9			X = Me	16.7
X = H	24.7			X = H	17.2

[a] Bordwell, F. G. *Acc. Chem. Res.*, 1988, *21*, 456; Bordwell, F. G., Harrelson, J. A. *Can. J. Chem.*, 1990, *68*, 1714; Zhang, X. M., Bordwell, F. G. *J. Org. Chem.*, 1993, *59*, 6456; [b] pK$_a$ = 18.2 for acetone enol

9.2.5 Carbon acids: nitriles, sulphones

Acid	pKa	Acid	pKa	Acid	pKa
Nitriles: CNCH$_2$X		X = 3-CN	18.7	NC–CH(X)–CN	
X = COPh	10.2	X = 3-CF$_3$	19.2	X = *p*ClC$_6$H$_4$	3.1
X = CN	11.1	X = 3-Br	19.4	X = Ph	4.2
X = SO$_2$Ph	12.0	X = 3-Cl	19.5	X = *p*OMeC$_6$H$_4$	5.6
X = CO$_2$Et	13.1	X = 3-F	20.0	X = NMe$_2$	12.2
X = H	31.3	X = 4-Cl	20.5	X = H	11.1
X = Me	32.5	X = 3-OMe	21.6	X = Me	12.4
X-C$_6$H$_4$CH$_2$CN		X = H	21.9	**Sulphones:**	
X = 4-NO$_2$	12.3	X = 4-F	22.2	PhSO$_2$CH$_2$CN	12.0
X = 4-CN	16.0	X = 4-Me	22.9	(MeSO$_2$)$_2$CH$_2$	15.0
X = 3-NO$_2$	18.1	X = 4-OMe	23.8	(EtSO$_2$)$_2$CHCH$_3$	16.7
X = 3-SO$_2$Ph	18.5	X = 4-Me$_2$N	24.6	CH$_3$SO$_2$CH$_3$	31.1

[a] Bordwell, F. G., Cheng, J-P., Bausch, M. J., Bares, J. E., *J. Phys. Org. Chem.*, 1988, *1*, 209

9.2.6 Carbon acids: fluorenes

Acid	pKa	Acid	pKa	Acid	pKa
9-X-fluorene		X = CONH$_2$	14.8	X = Et	22.3
X = NO$_2$	7.1	X = Ph	17.9	X = Me	22.3
X = CN	8.3	X = Cl	18.9	X = H	22.6
X = COPh	10.0	X = F	22.3	X = *i*-Pr	23.2
X = CO$_2$Me	10.3	X = Pr	22.2	X = *t*-Bu	24.4

9.2.7 Cationic acids: anilinium, ammonium, pyridinium ions

Base[a]	pK_a	Base[a]	pK_a	Base[a]	pK_a
ammonia	10.5	n-butylamine	11.12	4-OMe-aniline	5.08
methylamine	11.0	di-n-butylamine	10.0	4-Me-aniline	4.48
dimethylamine	10.3	tri-n-butylamine	8.4	aniline	3.82
trimethylamine	8.4			4-chloroaniline	2.86
ethylamine	10.7	Me$_4$-guanidine	13.2	3-chloroaniline	2.34
diethylamine	10.5	piperidine	10.85	3-cyanoaniline	1.36
triethylamine	9.0	pyrrolidine	11.06	N-Me-aniline	2.76
n-propylamine	10.7	pyridine	3.4	N,N-Me$_2$-aniline	2.51

[a] Kolthoff, I. M., Chantooni, M. K., Bhowmik, S. *J. Am. Chem. Soc.*, 1968, *90*, 23: Asghar, B. H. M., Crampton, M. R. *Org. Biomol. Chem.*, 2005, *3*, 3971: Mucci, A., Domain, R.; Benoit, R. L. *Can. J. Chem.*, 1980, *58*, 953

9.3 N,N-Dimethylformamide

9.3.1 Neutral acids

pK_a-values are summarized by Maran, F., Celadon, D.; Severin, M. G., Viannelo, E. *J. Am. Chem. Soc.*, 1991, *113*, 9320; the reference includes original data and literature values.

Acid		pK_a	Acid	pK_a	Acid[a]	pK_a
Carboxylic acid:			4-Cl, 3-NO$_2$	8.6	3-NO$_2$	15.4
Aliphatic:			3,5-Cl$_2$	8.8	4-NO$_2$	12.6
CHCl$_2$CO$_2$H		7.6	3,4-Cl$_2$	9.2	3-CF$_3$	15.7
CH$_2$ClCO$_2$H		10.2	2-NO$_2$	8.2	3-Cl	16.3
PhCH$_2$CO$_2$H		13.5	3-NO$_2$	9.2	4-Cl	16.8
CH$_3$CO$_2$H		13.5	4-NO$_2$	9.0	H	18.9
oxalic:	pK_{a1}	8.6	2-Cl	9.3		
	pK_{a2}	16.6	4-Cl	10.1	thiophenol	10.7
malonic:	pK_{a1}	7.9	3-Br	9.7	4-NO$_2$-thiophenol	6.3
	pK_{a2}	19.3	4-Br	10.5		
succinic:	pK_{a1}	10.2	H	11.0	**Anilide:**	
	pK_{a2}	17.3	3-Me	11.0	X = Br; Y = CN	16.4
glutaric:	pK_{a1}	11.3	3-OH	11.1	X = Br; Y = Cl	17.9
	pK_{a2}	15.6	4-Me	11.2	X = Br; Y = H	18.9
adipic:	pK_{a1}	12.2	3,4-Me$_2$	11.4	X = H; Y = CN	19.8
	pK_{a2}	15.7	4-NH$_2$	12.8	X = H; Y = Cl	21.4
o-phthalic:	pK_{a1}	6.7			X = H; Y = H	22.3
	pK_{a2}	16.5	**Phenol:**			
			2,6-(NO$_2$)$_2$	6.2	**Amide:**	
Benzoic:			2,4-(NO$_2$)$_2$	6.3	nicotinamide	22.5
2,6-(OH)$_2$		3.6	3-CF$_3$,4-NO$_2$	10.4	benzamide	23.9
2,4-(NO$_2$)$_2$		6.5	3,5-(NO$_2$)$_2$	11.3	formamide	24.3
2-OH		6.8	2-NO$_2$	12.2	2-pyrrolidinone	25.0
3,5-(NO$_2$)$_2$		7.4				

Additional dissociation constants for neutral acids in DMF, NMP and DMAC can be estimated from correlations with those in DMSO (Section 9.2) as follows:

$$pK_a(DMF) = 0.96 pK_a(DMSO) + 1.56$$

$$pK_a(NMP) = 0.99 pK_a(DMSO) + 1.08$$

$$pK_a(DMAC) = 1.0 pK_a(DMSO) + 0.1$$

9.3.2 Cationic acids

Amine[a]	pK_a	Amine[a]	pK_a
ammonia	9.45	tri-n-butylamine	8.57
dimethylamine	10.4	triethanolamine	7.6
diethylamine	10.4	pyridine	3.3
triethylamine	9.25	Me$_4$-guanidine	13.65

[a] Kolthoff, I. M., Chantooni, M. K., Smagowski, H. *Anal. Chem.*, 1970, *42*, 1622; Izutsu, K., Nakamura, T., Takizawa, K., Takeda, A. *Bull. Chem. Soc. Japan*, 1985, *58*, 455; Roletto, E., Vanni, A. *Talanta*, 1977, *24*, 73

9.4 Acetonitrile

9.4.1 Neutral acids

Acid[a]	pK_a	Acid[a]	pK_a	Acid[a]	pK_a
Carboxylic acid:		2,6–Cl$_2$	18.2	3–NO$_2$	24.6
Aliphatic:		3,4–(NO$_2$)$_2$	18.0	4–NO$_2$	21.7
CHCl$_2$CO$_2$H	16.4	2–NO$_2$	18.8	4–CN	23.7
CH$_2$ClCO$_2$H	19.7	3–NO$_2$	20.2	3.4–Cl$_2$	25.1
CH$_3$CO$_2$H	23.5	4–NO$_2$	20.7	3–CF$_3$	25.6
oxalic: pK_{a1}	14.5	3-CN	20.0	3-Cl	26.1
pK_{a2}	27.7	4-CN	19.9	4-Br	26.8
malonic: pK_{a1}	15.3	3-Cl	20.1	H	28.5
pK_{a2}	30.5	4-Cl	20.9		
succinic: pK_{a1}	17.6	3-Br	20.2	**Fluorene:**	
pK_{a2}	29.0				
glutaric: pK_{a1}	19.2	4-Br	20.2	X = CO$_2$Me	22.5
pK_{a2}	28.0	3-OMe	21.3	X = CN	21.4

adipic: pK_{a1}	20.4	H	21.5			
pK_{a2}	26.9	3-Me	21.5	**Malonitrile:**		

Malonitrile structure: NC–CHX–CN

azealic: pK_{a1}	20.8	4-Me	21.9	X = 4–MeC$_6$H$_4$	17.6	
pK_{a2}	24.8	4-OH	21.6	X = 3–CF$_3$C$_6$H$_4$	14.7	
o-phthalic: pK_{a1}	14.2			X = 4–NO$_2$C$_6$H$_4$	11.6	
pK_{a2}	29.8	**Phenol:**				
		2, 4, 6–(NO$_2$)$_3$	11.0	**Sulphonic acid:**		
Benzoic:		2, 4–(NO$_2$)$_2$	16.4	4-chlorobenzene	7.3	
2, 6–(NO$_2$)$_2$	16.2	3, 5–(NO$_2$)$_2$	21.3	3-nitrobenzene	6.8	
2, 4–(NO$_2$)$_2$	16.6	2–NO$_2$	22.9	4-nitrobenzene	6.7	
3, 5–(NO$_2$)$_2$	17.7					

[a] Coetzee, J. F., Padmanabhan, G. R. *J. Phys. Chem.*, 1965, *69*, 3193; Kolthoff, I. M., Chantooni, M.K. *J. Am. Chem. Soc.*, 1965, *87*, 4428; Kolthoff, I. M., Chantooni, M. K.; Bhowmik, S. *J. Amer. Chem. Soc.*, 1966, *88*, 5430; Kolthoff, I. M., Chantooni, M. K. *J. Phys. Chem.*, 1966, *70*, 856; Kolthoff, I. M., Chantooni, M. K. *J. Am. Chem. Soc.*, 1969, *91*, 4621; Chantooni, M. K., Kolthoff, I. M. *J. Phys. Chem.*, 1975, *79*, 1176; Chantooni, M. K., Kolthoff, I. M. *J. Phys. Chem.*, 1976, *80*, 1306; Kütt, A.; Leito, I.; Kaljurand, I.; Sooväli. L.; Vlasov, V.M.; Yagupolskii, L. M., Koppel, I. A. *J. Org. Chem.*, 2006, *71*, 2829

The dissociation constants show excellent correlations with those in water and DMSO:

(a) Water:
   carboxylic acids: $pK_a(MeCN) = 1.6 pK_a(H_2O) + 14.9$
   phenols: $pK_a(MeCN) = 1.8 pK_a(H_2O) + 9.6$
(b) DMSO:
   carboxylic acids and phenols: $pK_a(MeCN) = 1.00 pK_a(DMSO) + 10.5$
   carbon acids: $pK_a(MeCN) = 1.00 pK_a(DMSO) + 12.9$

9.4.2 Inorganic acids and miscellaneous

Acid[a]	pK_a	Acid[a]	pK_a	Acid[a]	pK_a
HClO$_4$	1.57	H$_2$SO$_4$	7.9	NH$_4^+$	16.5[b]
CF$_3$SO$_3$H	2.60	HCl	10.4	HSO$_4^-$	25.9[c]
FSO$_3$H	3.38	HNO$_3$	10.6		

[a] Kolthoff, I. M., Chantooni, M. K. *J. Chem.Eng Data.*, 1999, *44*, 124; [b] Section 9.4.3; [c] Kolthoff, I. M., Chantooni, M. K. *J. Am. Chem. Soc.*, 1968, *90*, 5962

9.4.3 Cationic acids: ammonium, anilinium, pyridinium ions

Base	pK_a	Base	pK_a	Base	pK_a
Amine[a]:		**Aniline**[b]:		2-hydroxy	8.3
ammonia	16.5	2–NO_2	4.80	3-cyano	8.0
methylamine	18.4	2,6–Cl_2	6.06	4-cyano	8.5
ethylamine	18.4	2,5–Cl_2	6.21	2-acetylo	9.6
n-propylamine	18.2	4–NO_2	6.22	3-bromo	9.5
n-butylamine	18.3	4–F–3–NO_2	7.67	3-chloro	10.0
t-butylamine	18.1	3–NO_2	7.68	3-acetylo	10.8
benzylamine	16.9	2–Cl	7.86	3-hydroxy	12.6
morpholine	16.6	4–CF_3	8.03	H	12.6
piperidine	18.9	2,4–F_2	8.39	3-methyl	13.7
pyrrolidine	19.6	4–Br	9.43	2-methyl	13.9
dimethylamine	18.7	2-Me	10.50	4-methyl	14.5
diethylamine	18.8	H	10.82	3-amino	14.4
di-n-butylamine	18.3	4-OMe	11.86	2-amino	14.7
trimethlyamine	17.6	N,N–Me_2	11.4	4-amino	18.4
triethylamine	18.5				
tri-n-proplyamine	18.1	**Pyridine**[b]:			
tri-n-butylamine	18.1	2-chloro	6.8		
DBU[c]	24.3	2-bromo	7.0		

[a] Coetzee, J. F., Padmanaban, G. R. *J. Am. Chem. Soc.*, 1968, *87*, 5005; [b] Kaljurand, I., Kütt, A., Sooväli, L., Mäemets, V., Leito, I., Koppel, I.A. *J. Org. Chem.*, 2005, *70*, 1019; Augustin-Nowacka, D., Makowski, M., Chmurzynski, L. *Anal. Chim. Acta*, 2000, *418*, 233; [c] DBU = diazabicycloundecane

9.4.4 Phosphazene bases

The phosphazene bases, the structures of which are listed below the Table, exhibit a wide range of pK_a-values.

Base[a,b]	pK_a	Base[a,b]	pK_a	Base[a,b]	pK_a
4–OMe$C_6H_4P_3$(dma)	32.0	HP_1(pyrr)	27.0	2–Cl$C_6H_4P_1$(pyrr)	20.2
PhP_3(dma)	31.5	t–BuP_1(dma)	27.0	4–$CF_3C_6H_4P_1$(pyrr)	20.2
2–Cl$C_6H_4P_3$(pyrr)$_6$Net$_3$	31.2	PhP_2(dma)	26.5	2–Cl$C_6H_4P_1$(dma)	18.6
4–$CF_3C_6H_4P_3$(pyrr)	30.5	2–Cl$C_6H_4P_2$(pyrr)	25.4	2,6–$Cl_2C_6H_3P_1$(pyrr)	18.6
2–Cl$C_6H_4P_3$(dma)$_6$Net$_3$	30.2	4–Nme$_2P_1$(pyrr)	23.9	2,5–$Cl_2C_6H_3P_1$(pyrr)	18.5
4–$CF_3C_6H_4P_3$(dma)	29.1	4–OmeP_1(pyrr)	23.1	4–$NO_2C_6H_4P_1$(pyrr)	18.5
EtP_1(pyrr)	28.9	PhP_1(pyrr)	22.3	2–NO_2,4–Cl$C_6H_3P_1$(pyrr)	17.7
t–BuP_1(pyrr)	28.4	PhP_1(dma)	21.3	2–NO_2,5–Cl$C_6H_3P_1$(pyrr)	17.3
4–Ome$C_6H_4P_2$(pyrr)	28.2	4–Br$C_6H_4P_1$(pyrr)	21.2	2,4–$(NO_2)_2C_6H_3P_1$(pyrr)	14.9
PhP_2(pyrr)	27.6	PhP_1(dma)$_2$Me	21.0	2,6–$(NO_2)_2C_6H_3P_1$(pyrr)	14.1
MeP_1(dma)	27.5	1–NapthP_1(pyrr)	20.6		

[a] Kaljurand, I., Kütt, A., Sooväli, L., Rodima, T., Mäemets, V., Leito, I., Koppel, I. A. *J. Org. Chem.*, 2005, *70*, 1019; [b] The structures of the phosphazene bases are listed below

Phosphazene base structures:

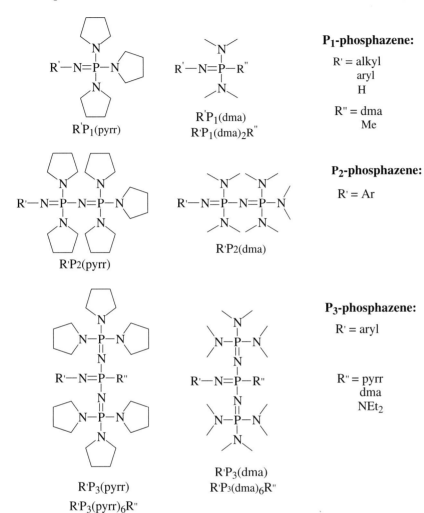

9.5 Tetrahydrofuran
9.5.1 Neutral acids

Benzoic acid[a]	pK_a	Phenol[b]	pK_a
2,4-dinitro	18.64	2,4,6-trinitro	11.84
3,5-dinitro	18.99	2,4-dinitro	16.94
2-nitro	21.10	4-nitro	21.13
4-nitro	21.16	2,4,6-trichloro	22.98
3,5-dichloro	21.64	3,5-dichloro	23.16
3-nitro	21.77	2,4-dichloro	23.50
2,6-dichloro	22.22	3-nitro	23.76
3-bromo	23.23	2-nitro	24.41
2-chloro	23.44	2,6-dichloro	25.10
4-chloro	23.88	2-chloro	26.30
H	25.11	4-chloro	26.80
3-methyl	25.34	4-bromo	27.30
2-methyl	25.39	H	29.23

[a] Barbosa, J., Barrón, D., Bosch, E., Rosés, M. *Anal. Chim. Acta*, 1992, *265*, 157; [b] Barrón, D., Barbosa, J. *Anal. Chim. Acta*, 2000, *403*, 339

9.5.2 Cationic acids

Base[a]	pK_a	Base[a]	pK_a	Base[a]	pK_a
N-EtP$_1$(tmg)[b]	32.6	**Amine:**		2-Cl	5.98
t-BuP$_1$(tmg)[b]	32.0	DBU[d]	19.1	3-NO$_2$	5.81
4-OMe-C$_6$H$_4$P$_3$(pyrr)	31.7	TMG[e]	17.8	3,4-Cl$_2$	5.33
4-OMe-C$_6$H$_4$P$_3$(dma)	30.5	pyrrolidone	15.6	2-NO$_2$	5.12
PhP$_4$(dma)[c]	29.8	triethylamine	13.7	4-NO$_2$	4.82
4-Br-C$_6$H$_4$P$_4$(pyrr)	29.7	propylamine	14.7	2,4-(NO$_2$)$_2$	4.61
EtP$_2$(pyrr)	29.4			3,5-Cl$_2$	4.47
EtP$_2$(dma)	28.1	**Aniline:**			
PhP$_3$(pyrr)	26.8	4-OMe	8.8	**Pyridine:**	
PhP$_3$(dma)	26.2	4-t-butyl	8.73	4-NMe$_2$	14.1
EtP$_1$(pyrr)	24.2	H	7.97	4-OMe	9.6
PhP$_2$(dma)	22.2	4-Cl	6.97	2-Me	8.6
PhP$_1$(dma)	17.8	3-Cl	6.38	H	8.25
4-NO$_2$C$_6$H$_4$P$_1$(pyrr)	15.8	4-Br	6.2		

[a] Garrido, G., Koort, E., Ràfols, C., Bosch, E., Rodima, T., Leito, I., Rosés, M. *J. Org. Chem.*, 2006, *71*, 9062: reference includes a more comprehensive set of phosphazene bases (structures are illustrated in Section 9.3.3); [b] tmg denotes N, N, N', N'-tetramethylguanidine radical; [c] P$_4$ denotes R$_3$P = N − P(= NR')(= NR$_3$)N = PR$_3$; [d] DBU = diazabicycloundecane; [e] TMG = N, N, N', N'-tetramethylguanidine

Index

Acceptor Number (AN) 8, 33–4, 36, 100, 114, 119
acetamide 72–3
acetic acid 1, 12, 19, 34–5, 49–50, 54, 69, 120–6; *see also* carboxylic acids
acetone 108–111
　acid dissociation of 86, 94, 126; *see also* carbon acids
acetonitrile 100–108, 118–20, 136–9
　acetonitrile-water 37–7, 69
　acid dissociation of 85–6, 94, 134; *see also* carbon acids
acetophenone 126; *see also* carbon acids
acid-base indicators 50–1, 74
acidity function, H_o 4
activity coefficients 41–3, 122–4, 125, 126
　solvent transfer 23, 24, 34, 35
adipic acid 80; *see also* dicarboxylic acids
alcohol-water 68–71, 120–4
alcohols:
　dissociation constants in dimethylsulfoxide 83–4, 132
amides 84, 133, 135
amidines 17, 113
amines 17, 64–5, 69, 86–7, 92–3, 94–5, 107–108, 113, 131, 135, 138, 140
　homohydrogen-bond formation by 45, 105, 125
amino-acids 71, 87–9
ammonia, liquid 93–5
anilines 64–5, 69–70, 73, 86–7, 106–107, 113, 114, 120–1, 131, 135, 138, 140
　ionization of neutral 84, 133
anthranilic acid 1–2
Atherton, J.H. 14, 46
autoprotolysis constant, *see* autoionization constant
autoionization constant 3, 53–4, 59–60, 76

Barrón, D., *see* Bosch, E.
Bates, R.G. 61
Bell, R.P. 15
benzoic acid 24, 74, 110, 113; *see also* carboxylic acids
Bhowmik, S. *see* Kolthoff, I.M.

Bordwell, F.G. 50, 76, 77, 89, 96
Born–Harber cycle 21, 25
Bosch, E. 61, 115
Brönsted, J.N. 10
t-butanol 66–8

calixarenes 53
carbon acids 3, 17–19, 85–7, 94, 104–5, 126, 133–4, 136
carbon tetrachloride 127–8
carboxylic acids 21, 61–3, 66–8, 72–4, 78–81, 101–4, 113–14, 130–1, 132, 135, 136, 140
　homohydrogen-bond formation by 44–5, 77, 103, 125, 126
Carpenter, K.J., *see* Atherton, J.H.
Chantooni, M.K., see Kolthoff, I.M.
chlorobenzene 127
chloroform 127–8
Coetzee, J.F. 109
crystallization 2, 13
Cunningham, I.D. 53

Davies equation 42, 48, 61
Debye–Hückel limiting law 42
dicarboxylic acids 62–3, 79–81, 91, 103, 130, 132, 135
dielectric constant 6, 60, 72, 99, 108, 109, 128
N,N-dimethylacetamide 89, 91–2, 92–3
N,N-dimethylformamide 89–91, 92–3, 135–6
　dimethylformamide-water 69
dimethylsulfoxide 3, 77–89, 118–20, 126, 132–5
　autoionization 3, 76
dimsyl ion 52, 84, 85
dioxane-water 69
dipole moment 7
distribution coefficients 25
Donor Number (DN) 7, 8, 33–4, 36, 93, 96, 100, 109, 114

Eigen, M. 99
enols 86, 112, 126

Index

esters 18, 133
$E_T(30)$ value 8
ethanol 66–8
    ethanol-water 60, 69–70, 120; *see also* alcohol-water

fluorenes 17, 112, 134, 136
formamide 72–3
formic acid (solvent) 75
free energies of transfer 21–2, 23–7, 62, 79, 82, 102, 108
fumaric, acid 92

glass electrode 40, 47–50, 53, 68, 111
glutamic acid 13–14
glutaric acid 81; *see also* dicarboxylic acids
glycine 88–9; *see also* amino acids
Grunwald, E. 49, 68
guanidines 17
Gutmann, V. 7, 8

Hammett equation 96, 118
Henry's Law 25
heptane 128
high-performance liquid chromatography (HPLC) 5
homoconjugation, *see* homohydrogen-bond formation
homohydrogen-bond formation 44–7, 55–7, 72, 77, 80, 83, 103
hydration of ions 27–8
hydrobromic acid 68, 133
hydrochloric acid 68, 133, 137
hydrogen-bond basicity, β 8, 36, 109, 114
hydrogen-bond acidity, α 8
hydrogen bonding 5, 8, 28, 35, 59, 80, 95
    intramolecular 79, 81, 91–2, 103
hydrogen electrode 39–40, 53

ion-pair acidity 111–13, 128
ion-pair basicity 113
ion-pair formation 43–4, 54–5, 100, 108, 111, 122–5, 127–8
ion solvation 27–35
ionic product, *see* autoionization constant
Izutsu, K. 77

ketones 18, 85, 133; *see also* carbon acids
Kolthoff, I.M. 47, 57, 77, 81
Koppel, I.A., *see* Leito, I.

Leito, I. 100, 105
Lewis basicity scale 7
Lowry, T.M., *see* Brönsted, J.N.

maleic acid 62–3, 92
malonic acid 62–3, 79–80, 81, 103; *see also* dicarboxylic acids
methanol 60–6, 118, 130–1
    autoionization constant 53–4, 59
    dissociation constant in dimethylsulfoxide 83–4
    methanol-water 49–50, 59, 69, 70; *see also* alcohol-water
methyl *iso*-butyl ketone 108–11
N-methylpyrolidin-2-one 1, 89–90, 92–3
mixed-aqueous solvents 120–4; *see also* acetonitrile-water

nitroalkanes 18, 85, 133–4; *see also* carbon acids
nitrobenzene 108–11
non-electrolyte solvation 32–3

pH-scales:
    aqueous 39–41
    non-aqueous 41
phenols 21, 63–4, 66–7, 81–3, 93–4, 101–4, 113–14, 118–19, 132, 135, 137, 140
    homohydrogen-bond formation by 44–5, 67, 83, 103
phosphazene bases 3, 105, 113, 114, 126, 128, 138–9, 140
picric acid 45, 82, 101, 113
P_n- bases, *see* phosphazene bases
polar aprotic solvents 5–6
preferential solvation 31–2, 120–21
i-propanol 66–9
propionamide, N-methyl 72, 73
propylene carbonate 108–11
protic solvents 5–6
pyridines 65–6, 86–7, 105–6, 113, 114, 131, 135, 138, 140

Reichardt, C. 6
Rosés, M. *see* Bosch, E.

salt formation 70–1, 120–7
Schwesinger, R. 52
selective solvation, *see* preferential solvation
sodium chloride 25–6, 36–7, 54
solubility measurements 23–4, 25–6, 46, 55

Index

Stewart, R. 15, 84
Streitwieser, A. 111
substituent effects 62, 79, 96, 115, 118–20
succinic acid 81; *see also* dicarboxylic acids
sulfolane 108–11

Taft equation 96
tetrahydrofuran 111–15, 119, 128, 140
  tetrahydrofuran-water 69

thiophenols 82–3, 136
triethylamine 1, 19, 125–6; *see also* amines

vapour pressure measurements 24

water:
  dissociation constant in dimethylsulfoxide 83–4
zwitter-ion 1, 13, 14, 71, 88–9